Advance praise for

Equity and Information
Communication Technology (ICT) in Education

"This book is a must for all educators. Developments in communication technology are having a profound effect on education. What has become increasingly important is to ensure that the advances facilitate efforts to meet disadvantaged groups—to close the widening information gaps. Professor Anderson does a masterly job in weaving innovations in the educational uses of ICT with research and evaluation (including the important work he and his team have done) to address equity issues. As the book rightly points out, empowering disadvantaged groups is not just about developing technical expertise but also developing the skills needed to select and apply appropriate software. Strengths of the book include the chapters on equitable access for girls and women, people with intellectual and physical disabilities, and closing the digital divide using low cost hardware and open source software. The book represents a thoughtful, state-of-the-art analysis and synthesis, and should serve as a handbook for practitioners and education policymakers."

*Emeritus Professor Colin Power, Chair of the Commonwealth Consortium
for Education; Editor-in-Chief of* Educational Research for Policy and Practice

Equity and Information Communication Technology (ICT) in Education

Colin Lankshear, Michele Knobel,
and Michael Peters
General Editors

Vol. 6

PETER LANG
New York • Washington, D.C./Baltimore • Bern
Frankfurt am Main • Berlin • Brussels • Vienna • Oxford

Neil Anderson

Equity and Information Communication Technology (ICT) in Education

PETER LANG
New York • Washington, D.C./Baltimore • Bern
Frankfurt am Main • Berlin • Brussels • Vienna • Oxford

Library of Congress Cataloging-in-Publication Data

Anderson, Neil.
Equity and information communication technology (ICT) in education / Neil Anderson.
p. cm. — (New literacies and digital epistemologies; v. 6)
Includes bibliographical references and index.
1. Computer literacy. 2. Educational equalization.
3. Digital divide. I. Title.
QA76.9.C64A53 004—dc22 2008045040
ISBN 978-1-4331-0051-2 (hardcover)
ISBN 978-0-8204-5243-2 (paperback)
ISSN 1523-9543

Bibliographic information published by **Die Deutsche Bibliothek.**
Die Deutsche Bibliothek lists this publication in the "Deutsche
Nationalbibliografie"; detailed bibliographic data is available
on the Internet at http://dnb.ddb.de/.

Cover design by Clear Point Designs

© 2009 Peter Lang Publishing, Inc., New York
29 Broadway, 18th floor, New York, NY 10006
www.peterlang.com

Contents

. . .

Acknowledgments

To my immediate family, Chris, Ben, Jean and Eve, I extend my love and appreciation for your ongoing support that enables my work to continue and hopefully, make a difference. To my extended family, Reg, Thelma, Fred, Keith, Trevor, Cassandra and Warren, thank you for your lifelong support.

Colin Lankshear has provided strong long term support for my professional life, for which I am most grateful. Many thanks to key members of an academic team that I've been fortunate to be part of over the last four years with Lyn Courtney and Carolyn Timms, who have also made valuable contributions to this book. Jane Buschens also made a significant contribution to the assistive technologies section.

Other wonderful influences on my professional life have been Paul Burnett, Hitendra Pillay, Michele Knobel and my colleagues in the School of Education at James Cook University, Australia. I wish to acknowledge that some small sections of chapters have been modified from previous publications such as the section 'Mindstorms and mindtools aren't happening' from the *Journal of eLearning*; games and Indigenous culture from the European Conference on Games Based Learning;

and Jhai Foundation Project from the *Encyclopaedia of Developing Regional Communities with Information and Communication Technologies* and rural education using blogs and podcasts from the *Proceedings of the Australian Computers in Education Conference, 2008*. I would like to acknowledge the support of the Australian Research Council through two consecutive grants for research in the area of gender and ICT and our industry partner, Technology One, for their financial and staff support.

Introduction

Information communication technologies (ICT) permeate almost every facet of our daily business and have become one of the most important priorities for formal and informal education. Readers in developed countries would be familiar with the type of rhetoric presented in the first sentence and many would identify with the sentiments, since they would resonate with the everyday evidence of their lives. However, many people, particularly those in least developed countries, would not have any empathy with the ICT driven lives common in developed countries. Nor would they always have any expertise in use of ICT or understanding of the benefits that ICT affords to people living in developed and developing countries. Nevertheless, very few people can escape the effects of globalization fuelled by new and emerging technologies.

ICT has become a key driver in culture, economics, politics and education with profound effects on all countries which in turn affect people in the most remote and least developed areas, even if they are not directly using the technologies. Culture, economics, politics and education are all intertwined, and the impact of ICT in one area often has flow on effects to other areas. For example, the level of ICT infrastructure a country builds up, directly affects the level of direct foreign investment which, in turn, affects the amount of knowledge and technology transfer. Knowledge transfer and technology transfer affect the willingness of foreign investors and the local government to provide education and training, influencing the likelihood of attracting and retaining expertise from various countries.

Another example of flow on effects is the original development of the Internet for the United States military and then the rapid spread to government, business and public use and its ubiquitous presence in education within developed and developing countries. It is difficult to determine the exact current level of spending on ICT by the U.S. military, but the Department of Defense Fiscal Year 2009 Budget Request showed one budget line as U.S. $68.5 billion dollars for communication and mission support systems and another one of $11.5 billion for science and technology. The Australian Defense Review in 2007 emphasized the importance of ICT and recommended a substantial ICT reform agenda. Israel's first high tech exports were produced by the Department of Defense—a trend that has continued over time, and New Zealand, a relatively small country spends $20 million per year on ICT for the military. The governments of these countries believe that being at the forefront of ICT development provides a strong measure of protection for their citizens and that lagging behind means that they are vulnerable. People with access to Internet or television services are only too familiar with the military advantage of technology enhanced weapons and military communications. Likewise, in the area of developing sustainable systems for environment protection, ICT has become the key driver, albeit with much smaller budgets than the military. In the arena of democracy, ICT via e-voting and online debate/campaigning is seen to be a solution to the lagging participation rates in voting, which has been identified as a threat to the democratic process.

This book presents many examples of the advantages and opportunities afforded by new technologies and examines innovative ideas that foster equitable access to ICT and the essential education and training that enables powerful use of new technologies. Education needs to go further than merely providing the technical expertise to operate software but also needs to develop the skills that all people need to be able to choose the most appropriate software and hardware tools for specific needs.

While some countries invest enormous amounts of money in research and development of ICT, and recognize the importance of education and training to reap the benefits associated with technology, many citizens in other countries are left wishing they had any form of electronic communication. Part 4 outlines a case study in Laos where the local villages requested access to the Internet and voice over IP (VOIP) telephone access as a higher priority than mains electricity. Providing these services without mains electricity was challenging but, in the end, achievable.

Provision of equitable access to information communication technologies along with associated education and training has become an important United Nations goal. This book provides information and seeks to stimulate discussion on many of the important aspects of ICT in relation to equity and education. Recognising the interconnected nature of equity issues with ICT in education, the book follows a structure that uses broad parts with sections rather than chapters. Each part provides evidence from the current and past literature but, where possible, also provides details of original research. The parts of the book include gender and ICT; intellectual disability and ICT; assistive technology; digital divide—including socioeconomic factors and rurality; software and hardware developments and finally knowledge and technology transfer.

Part One examines one of the most important current equity issues involving ICT and formal participation in education—the negative mind-set towards ICT of many female students in the United States, Australasia, Europe, Israel and some Asian countries such as Singapore. This mind-set has led to under-representation of girls in high school and university higher level courses with resultant negative effects on the ICT industry and the products and advances emanating from that industry. This development has been examined by many researchers in Western and Asian countries, and addressing the problem has become a high priority for large ICT companies such as IBM. The United States government has provided large research grants, and universities have set up substantial research centers such as The Center for Women and Information Technology at UMBC, Maryland. Likewise, other countries have established centers that engage in research, publication, events and courses along with advocacy. Recently, the APEC (Asia-Pacific Economic Community) international meeting included a major forum devoted to this problem. A team working with the author has engaged in long term research that has shed some new light on this problem and these illuminating results are presented in this section.

Part Two considers the use of ICT to enhance educational opportunities and outcomes for students with intellectual disabilities included in regular classrooms. In the United States the Individuals with Disabilities Act covers approximately 6.5 million students aged from 3 to 21 years with about three quarters of them being educated in inclusive school environments. An overview provides information on current legislation and the results of recent research in the United States and elsewhere. The next section presents a two year case study of how the infusion of ICT in students' everyday work led to the development of competence and the improvement of academic self-concept and peer acceptance. Reference is made to contemporary literature and research studies throughout the intervention discussion.

Part Three looks at a compelling example of the power of information communication technologies to provide an equitable educational environment for people with physical disabilities. The term 'assistive technologies' refers to technologies that assist people with disabilities to function better. Examples of assistive technology discussed in this section range from the latest research in the United States by Velliste, Perel, Spalding, Whitford and Schwartz (2008), which has enabled monkeys to control artificial limbs with their thoughts, to more mundane yet important examples that have made the difference between participation and nonparticipation in educational activities for many people. An extensive literature review gives a general overview of assistive technologies, examines emerging and future trends and then focuses on issues related to educators and assistive technology.

Part Four is devoted to a discussion on the digital divide, including socioeconomic disadvantage and rurality. The first section goes beyond a typical deficit approach to the digital divide that sometimes focuses only on access to ICT resources. This section extends the argument to making a case for 'particular' uses of ICT in education that make a positive difference to educational outcomes for students affected by poverty. The following section uses informative case studies to illustrate various ways that ICT can assist to alleviate poverty in developing and least developed countries. For example, in the Jhai Project currently being undertaken in Laos, the local villagers prioritised ICT as a pressing community need, even though mains electricity was not available. The case study reveals how computer use, Internet connectivity and VoIP (voice over IP technology) were enabled despite the challenges of geographical isolation and poverty. Other case studies from developing countries are presented to illustrate diverse ways of approaching the problem. A further case study from a developed country is also presented and is centred on the use of blogs and robotics by an elementary school teacher in an extremely geographically isolated location to enhance a classroom unit of work and to provide the students with electronic connectivity to other isolated students working on the same unit.

Part Five focuses on hardware and software developments and reviews the recent advantageous and powerful coupling of low cost hardware with Open Source Software. Nicholas Negroponte's and MIT's One Laptop Per Child (OLPC) program is reviewed, since it acted as a catalyst to a 're-think' of the production of ever increasing power in portable and desktop computers and the continual development of bloated and expensive software. Some researchers and writers have argued that the OLPC project became a commercial threat to hardware and software manufacturers, who then responded with their own version of low cost mobile computers. This turn of events promises to assist in the affordability of ICT resources. As previously discussed in other sections of the book—pure access to ICT resources is not by itself a means of overcoming disadvantage. A key factor in the potent use of ICT is the way that computers are used and the interaction of the participants through the available software and associated educational activities. With this in mind, the second section presents information about the rise of serious games and their potential to assist in the education of people with particular needs. It also presents an interesting pilot study to examine the use of games to promote an understanding of indigenous cultures.

Part Six examines a major area of research that is usually confined to the business and economics domain. This important area of research has been labeled 'knowledge transfer,' 'technology transfer' or 'knowledge exchange' and is of immense importance to international equity in the use of ICT in education. International knowledge transfer facilitated by ICT has the potential to improve conditions for people in all countries, especially developing and least developed countries. This section defines the important terms and concepts and then presents a report based on in-depth interviews with key players involved in international knowledge transfer. The responses provide insight into the factors that facilitate or hinder the international transfer of knowledge. The final section deals with one of the most important international relationships for knowledge transfer, between the United States—the world's richest economy—and China—the world's most populous country.

. . .

PART I

Gender and Information Communication Technologies

WITH CAROLYN TIMMS

Investigating the Attitudes of Girls Towards ICT

Information and communication technologies are not a panacea or magic formula, but they can improve the lives of everyone on this planet. However, while technology will shape the future, it is people who shape technology, and decide what it could and should be used for. These new technologies should, therefore, be embraced, while recognizing that this is an endeavour that transcends technology. Speech by Kofi Annan, UN Secretary-General to the World Summit on Information Technology (2003).

Introduction

In 1997, when Tracey Camp observed that there were ever plummeting numbers of women achieving bachelor's degrees in computer science and subsequently going on to work within the Information Communication Technology (ICT) industry, the analogy she used was of an 'incredible shrinking pipeline.' At the time, 28% of ICT graduates in the United States (U.S.) were women. The years following the publication of Camp's influential paper have seen the numbers of women graduates in ICT fall even further, for example, Zweben (2005) cited 17% female ICT graduates for 2004 (2005). The same story has been repeated in Australia, with the Department of Communications, Information Technology and the Arts (DCITA) (2005) reporting at the national level, that 80% of new students enrolling in undergraduate ICT courses in 2003 were male. At a state level it was reported that women make up 16.7% of ICT professionals in Queensland (Queensland Government Office for Women, 2005).

To make matters worse, this decline is embedded within an overall decline in interest on the part of students (male and female) in undertaking computer science as a major (Vegso, 2005). A recent Department of Education, Science and Training (DEST) (2006) report noted that in Australia only 6.4% of all university enrolments were in ICT, but alarmingly, within that figure, 2.4% of all female university student enrolments were in ICT. Given that the Australian Bureau of Statistics (ABS) (2006) reported that 53% of higher education students were female, it could well be that falling overall percentages mask an even greater decline in the numbers of women undertaking ICT courses. Therefore, regrettably, Camp's shrinking pipeline analogy is apt, the trickle of women undertaking undergraduate courses necessarily translates to ever reducing numbers of women undertaking graduate degrees and taking up employment as ICT professionals.

The inevitable response in some quarters might well be, "So?"

The Ubiquitous Nature of the ICT Industry in Modern Society

Even the most fleeting glance will reveal an ICT industry at the raw stages of growth, hydra like, highly dynamic, globally competitive and innovation ready (Moore, Griffiths & Richardson, 2005). Modern Western societies are remarkably accommodating to constant change in technology. Irrespective of individual views as to whether this is or is not a good thing, computer literacy has shifted from the dark and deeply mysterious realm of specialists to an important survival skill for all. Information Communication Technologies are becoming part of the fabric of most aspects of global business and social action. Importantly and specifically, ICT:

> Provides the communication infrastructure and the means to access and store data and to manipulate it to provide information critical to business, government, research education and community operations. Using ICT effectively has meant complementary changes in organisations that yielded efficiency gains as well as enhanced capabilities of outputs. ICT is critical to the achievement of Australia's national goals, for example, economic growth, national security, dealing with demographic change, environmental management, education and health (DCITA, 2005, Attachment A, p. 1).

The primary concern, echoed in other Western nations, is therefore that reducing enrolments in ICT courses will lead to shortages of ICT professionals at a time when their contribution to society is most needed. Diverse reports and projections indicate growing labour shortfalls for most world regions in ICT industries at professional levels. The demand for skilled and knowledgeable ICT professionals is growing faster than formal education can provide (Australian Bureau of Statistics, 2006; Zweben & Aspray, 2004). As industries, government departments, and business operations expand and seek to compete in a global marketplace, ICT professionals will become integral members of multi-disciplinary teams. Furthermore, ICT professionals working on teams with professionals from other disciplines will require more than technical knowledge. They will need to: understand how business works; have prowess in problem solving; possess good negotiation skills; have the ability to identify with customers; and project colleagues and accurately anticipate their needs. "These are skills that women are traditionally thought to have" (DCITA, 2005, p. 2).

Within this context, a large and growing mass of empirical research documents disproportionately low levels of participation by female students in ICT courses at all levels and within professional level ICT careers in Australia and many other countries (DCITA, 2005; Millar & Jagger, 2001; Newmarch, Taylor-Steele & Cumpston, 2000; e-skills United Kingdom (UK), 2006). Under these conditions, low female participation rates at professional levels of the ICT industry is simultaneously an equity issue, when viewed from the standpoint of female participation, and a pragmatic issue with potentially far-reaching implications for the competitive advantage of firms, regions, countries and economic blocs (Frieze, 2005; Oudshoorn, Rommes, & Stienstra, 2004; Woodfield, 2002).

The DICTA (2005) report observed, "a significant bottleneck to the supply of ICT skills is the education choices students make" (Attachment B, p. 5). During the middle years of high school, students make choices about what subjects they will take at senior level (for example, years 11 and 12 in Queensland, Australia). These choices then direct the student towards particular career pathways at tertiary level. Therefore, while the looming shortage of ICT professionals is more than a gender issue (Vegso, 2005), according to the American Association of University Women (AAUW, 2000), it is important that gender focussed research is conducted. This is not only for the purpose of achieving understanding of why girls and women in particular may not be attracted to ICT careers, but may provide indications as to why the field still fails to achieve a broad based community appeal, in spite of the all pervading nature of computer technology in modern Western societies.

> In some important ways, the computer culture would do well to catch up with the girls. In other words, girls are pointing to important deficits in the technology and the culture in which it is embedded that need to be integrated into our general thinking about computers and education. Indeed, girls' critiques resonate with the concerns of a much larger population of reticent users. The commission believes that girls' legitimate concerns should

focus our attention on changing the software, the way computer science is taught, and the goals we have for using computer technology (AAUW, 2000, p. ix)

To date, research into why girls are not attracted to ICT has been prolific, and so far a wide ranging and remarkably consistent array of reasons has been advanced from numerous sources.

Why Girls are Not Attracted to ICT— The "Usual Suspects

1. A masculine environment

The first argument is somewhat circular but nonetheless compelling. As many of the young men who choose advanced computing subjects at senior level have honed their technological skills as children playing computer games, a persuasive line of reasoning is advanced by researchers asserting that these games have been created by males for males and that young girls find them boring and repetitive. It has, therefore, been argued that the common childhood practice of computer game playing (almost invariably a 'boy thing') has consequently enabled boys to develop computer fluency (Adya & Kaiser, 2005; Agosto, 2004; Armstrong, 2005; Frenkel, 1990; Gorriz & Medina, 2000; Gurer & Camp, 2002; Lynn, Raphael, Olefsky & Bachen, 2003). The computer games argument notes this computer fluency and suggests that it then becomes 'assumed knowledge' when the student reaches computer classes in high school. Thus girls, by comparison, are disadvantaged and alienated because of the failure of electronic games to attract their interest in childhood, which has led to a comparative lack of fluency with technology.

Furthermore, according to a number of researchers, girls are intimidated by the male environment of the computer classroom and are very conscious of their lack of technological fluency (Cisco Systems, 2002; Gurer & Camp, 2002; West & Ross, 2002). Extending the theme of intimidation within the computer classroom, Cohoon (2003) indicated that at tertiary level women who do choose an ICT major were overwhelmed by a lack of peer support. Gurer and Camp (2002, p. 122) noted "quite a few first year computing courses assume a certain level of knowledge that, through no fault of their own, many female students have not yet obtained."

The usual suspects. 2. Social and Biological Programming

Social programming is high on the list of reasons advanced by other researchers. McNair, Kirova-Petrova and Bhargava (2001) noted that advertisements for technology showed a gender differential in how they portrayed computer users; men and boys were portrayed in active and professional roles, whereas girls were portrayed passively as pretty or seductive. They further observed:

> To summarize, there are too few females modeling competent computer use in schools, inadequate opportunities for girls to get together to explore computers, and an imbalance of software titles such that most choices are designed for more typical boy interests. These experiences influence females' later confidence in computer activities, their choice of computer courses, the amount of time they choose to spend working with computers, and their persistence when presented with barriers to computer use. (McNair et al., 2001, p. 52)

Other researchers advance sociological or biological reasons. "Society has socially conditioned most men to interrupt and insert their opinion in a meeting, while women will wait for an opening in the conversation, which often never happens" (Gurer & Camp, 2002, p. 123), and this is said to compound intimidation of girls in male dominated classrooms. However, De Palma (2001, 2006) advanced biological reasons, not for the reluctance of girls to undertake computing, but for the attraction that current computer curricula hold for boys. His assertion was that many young men are drawn to computer science by the 'tinker factor':

> They enjoy taking things apart and putting them back together. They like kits, gadgets and screwdrivers. They were the boys who set up the audio-visual equipment in high school 30 years ago, and who now 'man'—the choice of gender ID deliberate—the school's computer network. They are fascinated with anything that moves, especially if it has wheels or wings, and, crucially, is not alive. (De Palma, 2001, p. 27)

De Palma's point was not that boys were biologically more suited to computer science, but that curricula are generated by males who like to 'tinker' and consequently more directed by and sympathetic to the 'tinker factor.' This was supported by Lane (2005), who also suggested that females were disadvantaged by the fact that their learning styles differ from those of males. Broos (2005) acknowledged that males and females have similar attitudes in regard to the perceived usefulness of computers, however, women are more likely to experience lower self-confidence about their ability to interact with one. It is thought that this lower self-confidence in the face of 'assumed knowledge' (Gurer & Camp, 2002) and lack of mentors (Cohoon, 2003) translates, inevitably, into a stressful experience for girls. The AAUW (2000) observed:

> We found that girls observe and describe strong gender differences but do not have a language with which to talk about them. The result is that girls are likely to express bewilderment and confusion about how they are different in their attitudes and abilities than boys. In girls' efforts to find a perspective from which to talk about gender differences, they often position themselves as morally or socially more evolved than boys who, they tell us, enjoy "taking things apart" and interacting with machines. (p. 8)

Many authors observed gender differences in self-efficacy (Bandura, 1997) or confidence with technology, including individual assessment of ability, which in many cases did not correspond to reality, for example:

> There is strong evidence, for example that women, even though they perform at the same levels, have less confidence in their abilities and individual accomplishments than men. Women are often less aggressive than male students in promoting themselves, attempting new or challenging activities, and pursuing awards or fellowships. (Cuny & Aspray, 2002, p. 168)

Therefore, not surprisingly, it was observed by a number of researchers that those girls who chose computing careers were invariably encouraged by influential people in their environment such as empathetic teachers, parents or peers (Adya & Kaiser, 2005; Gates, 2002; Turner, Bernt & Pecora, 2002; von Hellens, Nielsen & Trauth, 2001).

Beyond the usual suspects

According to De Palma (2005; 2006), computer science, unlike most other sciences such as mathematics and engineering, is not a fixed body of knowledge which can be imparted and learnt but a discipline dependent on informal knowledge gained from hours of trial and error at a computer keyboard (tinkering). It was advanced in the previous section that the 'tinker factor' is perhaps an important aspect when considering possible blockages in the shrinking pipeline analogy. Nevertheless, however compelling many of the arguments in the previous section are, they fail to address the fact that women are now achieving success in many other previously male dominated (and therefore ostensibly 'alienating to women') fields. The traditionally male preserves of medicine, science and law have been seen as a challenge to be conquered and a source of fulfilment for the women concerned. A recent ABS (2006) report noted that 53% of Australian students between 15 and 64 years of age were female. Another report from DEST (2006) reported that of 15,579 students who completed Natural and Physical Science courses in 2004, 8,702 (56%) were female. In addition, Moran (2004, 26 November) reported that more than half of law graduates in Australia were female. These are career fields which were previously dominated by men. It is, therefore, not sufficient an explanation to say that women are intimidated by male dominated classrooms, or by superior inherent male dexterity with matters technological.

The Current Research

Background

One of the authors of this chapter (Anderson, 2003) was involved in the organisation of the "I-Star" program for senior high school students between 2001 and 2004. This was an Australian State Government funded initiative designed to provide girls with rich and intensive opportunities to interact with technology. This program operated as a series of three day weekend camps with up to 50 high school girls each year. The camps were held in far north Queensland; however, girls attended from urban, regional and rural areas all over the state, along with interested teachers and other members from their communities. Those running the workshops, selected for their 'switched on' qualities, introduced participants to aspects of ICT such as multimedia animation, web design and video editing. The program was highly successful and participants were enthusiastic and keen about their newfound familiarity with

ICT, as an accessible, student friendly and creative experience. However, the girls also drew stark contrasts between this program and school computing subjects. Feedback invariably referred to the "boring" nature of their previous experience with ICT. Repetition of this comment on feedback forms over the years 'I-Star' program operated provided inspiration for the research program reported here.

The Girls and ICT Research Project

This section focuses on outlining the major findings of a research project conducted at James Cook University (JCU), Cairns Campus in Northern Australia. The purpose of the research was to investigate factors associated with low participation rates by females in education pathways leading to professional level information and communications technology (ICT) professions. The project was funded by the Australian Research Council (ARC) under its Linkage Grants Scheme and was conducted in collaboration with state education authority—Education Queensland (EQ) and ICT industry partner Technology One. The research was conducted over three years, and the research design included four stages:

a. A pilot survey of high school girls (years 11 and 12) which sought to determine major factors influencing the decisions of girls to avoid subjects leading to professional ICT career pathways

b. A major survey (pen and paper) of year 11 and 12 girls in Queensland

c. Follow-up focus groups with year 11 and 12 girls in Queensland

d. An online survey of women employed within the ICT industry or working in an ICT capacity within another industry in Australia

Detailed accounts of methodology and major findings can be found in a number of academic publications that have emanated from the research team during the course of the project. Anderson, Klein and Lankshear (2005) outlined preliminary work and rationale of the research. Findings from the high school survey phase were published in a number of quarters (Anderson, Lankshear, Courtney & Timms, 2006; Anderson, Lankshear, Timms & Courtney, in press; Courtney, Timms & Anderson, 2006; Timms, Courtney & Anderson, 2006. In addition to academic publications, the research has stimulated interest in the corporate sector, two examples of this interest include a nine page feature in *Business Review Weekly*, 'EXI.T. Why Women are Shunning the Technology Industry' (Walters, 2006) and a technology feature which appeared in both the *Sydney Morning Herald* and Melbourne's *The Age* in November 2006 (Millar, 2006, November 14). As focus group findings have not as yet been published, the current chapter will provide an outlet for them, within the context of the project as a whole. Finally, findings from the fourth part of the research, a survey of women working in the ICT industry, have been published in journals in the United Kingdom and the United States (Courtney, Lankshear, Timms & Anderson, 2008; Timms, Lankshear, Anderson & Courtney, 2008).

Pilot study

Details of the methodology of the pilot study, conducted in 2004, with 171 high school girl respondents, are outlined in Anderson, Lankshear, Timms and Courtney (2007). Respondents who had not chosen to undertake advanced computing subjects in their senior years at high school (Non Takers) expressed opposite opinions to those students who had chosen to take advanced computing (Takers). Non Takers (n =148) of the two advanced computing subjects in Queensland (Information Processing Technology [IPT] or Information Technology Systems [ITS]) strongly endorsed: "the subjects are boring" (45%); "I am not interested in computers" (50%); and "The subject would not be helpful for me in my chosen career path after school" (49%). Interestingly the pilot study found that the same reasons (albeit in the positive rather than the negative format) came from Takers as reasons why they had chosen to undertake advanced computing at a senior level in high school. Of 23 Takers, 78% endorsed "The subjects are interesting"; 61% agreed with, "I am very interested in computers; and 49% agreed that "the subject will be helpful in my chosen career after school."

The obvious dichotomy of opinion between Takers and Non Takers was of particular interest to researchers, who were interested to establish where the clearly polarised opinions come from. In addition, the fact that gender issues, so widely cited in the literature, such as intimidation of female students by males (Gurer & Camp, 2002; West & Ross, 2002), being alienated by the proliferation of computer games designed for a male audience (Gorriz & Medina, 2000; Millar & Jaggar, 2001) and consequent lack of familiarity and skill with computers (Adya & Kaiser, 2005), did not feature in responses of student respondents to the pilot survey but was also noteworthy. This was particularly interesting in the case of the Takers, whose upbeat attitude towards their advanced computing subjects was diametrically opposed to previous research findings and the views of Non Takers. The pilot study therefore provided the framework for the rest of the project by confirming that:

» There were multiple factors involved in student decision making
» Different groups of girls within the same school community had divergent perceptions of the same subjects

The high school survey

Detailed methodology of the high school survey is provided elsewhere (Anderson, Lankshear, Timms & Courtney, 2007; Courtney, Timms & Anderson, 2006; Timms, Courtney & Anderson, 2006). Questions for the survey were refined in light of pilot study findings and extra questions pertaining to respondents' participation in special ICT events (an example being the "I-Star" program previously described, but there are a number of other such programs operating in Queensland, for example: GIDGITS (Girls Into Doing Great Information Technology Society, 2007). An American example of such a program is the "Roadshow" (Frieze, 2005), where women ICT staff and students from Carnegie Mellon University (U.S.) initiated a program touring local high schools with engaging ICT activities, thereby providing students not only with examples of ICT's flexibility but also interaction with keen and knowledgeable women mentors.

The survey also provided extra space for respondents' comments. Although a total of 5,863 surveys were distributed to 31 schools in Queensland, only 1,452 completed surveys were returned to researchers from 26 schools (25% response rate). Unfortunately many schools reported research fatigue, and researchers were conscious of the need to maintain the goodwill of long suffering school staff who not only administered the survey, but who reminded students to return completed consent forms. However, before half of the completed surveys were received, trends were noticed in the data which did not change as more surveys came in. One of these was the proportion of Takers (approximately 9%) to Non Takers; another trend which remained enduring, once established, was the pattern of responses to the questions. This indicated to researchers that had more responses been available, it is unlikely that findings would have changed substantially.

Respondents to the survey were year 11 and 12 girls at Government schools (GS) and Non Government schools (NGS) in Queensland. Of a total of 1452 respondents, 1322 were Non Takers and 130 were Takers. Further details of respondents may be found in Table 1.

Table 1. Breakdown of participant details

	Takers of IPT/ ITS (or both)	%	Non Takers of IPT/ITS	%	Total
Year 11	80	5.6	677	46.6	757
Year 12	47	3.2	632	43.5	679
Year unknown	3	.2	13	.9	16
Totals	130	9	1322	91	1452

Data analysis

Table 2 reports the findings of Mann-Whitney U test comparisons between Takers and Non Takers on questions outlined in Table 2. Statistically significant differences were found between the two groups on four items: 'the subjects are interesting,' 'I am very interested in computers,' 'the subject will be helpful to me in my chosen career path after school' and 'it suited my timetable.' These are also displayed in Figure 1, where a horizontal line is drawn indicating the midpoint "Neither Disagree or Agree line."

It is noted that of the items that reached significance, three attracted more agreement from Takers: "The subjects are interesting"; "I am very interested in computers";

"the subject will be useful in my chosen career"; and one attracted more agreement from Takers, "ICT subjects suited my timetable." The last one indicating that Non Takers didn't regard timetable issues as relevant to their choice not to take advanced computing subjects at a senior level. Two items showed no significant difference between the groups: 'I have a lot of experience using computers' and 'I thought it would help my OP (Overall Position) score' ('OP' refers to Queensland's tertiary entrance system). In other words, both Takers and Non Takers believed the subjects would help their OP score, and likewise both groups felt they had experience in using computers. Whereas the Takers chose the subjects, the Non Takers did not. Conversely, the reasons for the differences in the choices seemingly have to do with interest in computers, the subjects, timetable issues and beliefs about the subject in relationship to future careers.

Table 2. Mann-Whitney U comparisons of relationships between Takers and Non Takers and study variables

	Takers			Non Takers			U Value	Significance
	N	Mean	SD	N	Mean	SD		
The subjects are interesting	130	3.89	.99	1314	2.60	1.26	37402.0	.000***
I am very interested in computers	130	3.76	1.03	1314	2.85	1.31	51609.0	.000***
I have a lot of experience using computers	129	3.67	.95	1315	3.51	1.17	79032.0	.186 (ns)
I thought it would help my OP score	130	3.38	1.03	1308	3.55	1.10	78626.5	.142 (ns)
The subject will be helpful in my chosen career	130	3.78	1.06	1314	2.52	1.24	39046.5	.000***
It suited my timetable	130	3.15	1.16	1306	3.49	1.18	72152.0	.003**

Note. Discrepancies in participant numbers reflect missing responses for these items.
** p < .01, *** p < .001

Figure 1. Comparison of Takers and Non Takers on statistically significant study variables

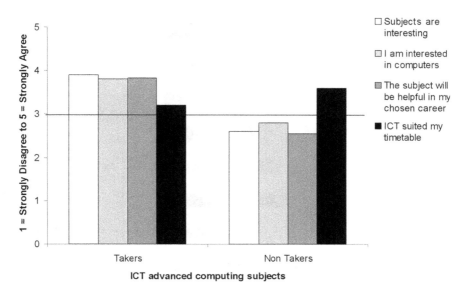

Attitudes of respondents to the 'usual suspects'

Figure 2 outlines attitudes of Non Takers (n = 1322) in regard to 'usual suspects' within the literature. Respondents in this survey tended to agree with, "The subjects will not be useful in my chosen career path," "I am not interested in computers" and "The subjects are boring" shown in black in Figure 2. Questions such as: "I don't have much experience in using computers"; "I did not think it would help my OP (university entry score)," "I don't think I would like the way the subjects are taught" and "did not suit my timetable" did not demonstrate a clear agreement pattern. It is noted that many Non Takers, who represented a large group of high school girls, agreed with these statements, and consequently these statements are represented by grey on the graph. The most definitive inclinations towards disagreement were with statements representing the 'usual suspects' within the literature. These included: "there are too many boys in these subjects; "I don't have a computer at home"; "I was discouraged by friends"; "I was discouraged by a family member"; "I was discouraged by a teacher or careers counsellor"; "there is not much content for girls"; "I think you need to be good at maths"; and "I would worry about what my friends would think." Table 3 outlines frequencies of responses to each of the various items.

Figure 2. Non Takers' (n = 1322) responses to many of the usual reasons advanced for the nonparticipation of girls in advanced computing classes

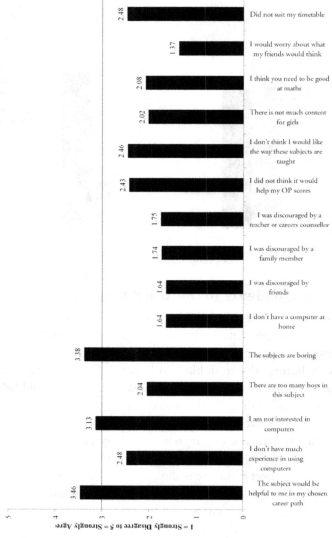

1 = Strongly Disagree to 5 = Strongly Agree

Comparison between two surveys in the research project: For purposes of comparison, the comment 'ICT is boring' was also inserted in the 'Women in ICT' survey conducted in the fourth stage of the research project. Figure 3 demonstrates a comparison of responses from Takers (n = 130), Non Takers (n = 1322) and women who work in the ICT industry (n = 270). What is particularly striking about Figure 3 is that the responses of the Takers echo those of women in the ICT industry, whereas Non Takers' responses diverge strongly.

A significant analysis of variance (ANOVA) (F (2, 1719) = 188.11, p < .001) confirmed that there was a significant difference between Takers and Non Takers on tatement Figure 3, Scheffe post hoc tests confirmed that the small difference between Women in the ICT industry and Takers was not significant. Actual frequencies and percentages for each category of response is presented in Table 2 and means and standard deviations for these groups are provided in Table 3.

Table 3. Frequencies of responses of Takers, Non Takers and Women in the ICT Industry to "ICT is boring"

	Women in ICT Industry		Girls Takers		Girls Non Takers		Total	
	N	%	N	%	N	%	N	%
Strongly disagree	84	31.1	34	26.2	135	10.3	253	14.8
Disagree	132	48.9	67	51.5	163	12.4	362	21.1
Neither agree or disagree	32	11.9	14	10.8	373	28.4	419	24.4
Agree	13	4.8	11	8.5	329	25.0	353	20.6
Strongly agree	9	3.3	4	3.1	314	23.9	327	19.1
Total	270	100	130	100	1314	100	1714	100

Note. Discrepancies in numbers reflect some missing values.

Figure 3. Comparison of Takers, Non Takers and Women in the ICT Industry to "ICT is boring"

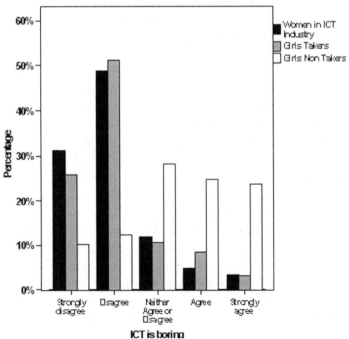

Table 4. Means and standard deviations of responses to "ICT is Boring"

	N	Mean	SD
Women in Industry	270	2.00	.96
Girls Takers	130	2.09	1.00
Girls Non Takers	1322	3.38	1.28
Total	1722	3.07	1.34

The purpose of this last comparison was to demonstrate that girls who were Takers of advanced computing subjects and who therefore have had experience in using technology had similar perceptions to women who work in the ICT industry and disagreed that ICT was boring. However, Takers represented only 9% of respondents in the present survey. Their response posed a sharp contrast to the attitudes of Non Takers whose view clearly supported previous findings of the Victoria State Government (2001):

> The research found that most young people have an overwhelmingly negative perception of IT jobs as male-dominated, boring, solitary and low-level tasks that involve little more than sitting at a screen all day entering code or data. They believe that IT employees interact only with a computer, not with people. (p. 5).\

Views of women working in the ICT industry about their careers

Before continuing with discussion of the high school survey and focus group responses, the current section is devoted to outlining some results from the 'Women in ICT Industry' survey. The purpose of this is to demonstrate that perceptions of Non Takers towards ICT are largely based on enduring stereotypes with little foundation in reality. More extensive information about the Women in ICT Industry survey can be found in Anderson, Timms and Courtney (2006), Courtney et al. (in press) and Timms et al. (2008).

The online Australia-wide Women in ICT Industry Survey was active between May and October, 2006. Women employed as ICT professionals were invited to participate. With relatively few women working in the Australian ICT industry per se, ICT professionals were sought from two sources: the ICT industry itself, and other industry sectors that involve professionals working in ICT. Of the 272 women who responded to the Women in ICT Industry Survey, 178 identified themselves as working in the ICT industry, and the remaining 96 women worked as ICT professionals in other industries such as education or finance.

Many of the questions inserted into the Women in ICT Industry Survey were included to establish some understanding of respondents' real life experience in contrast to the widely held perceptions of Non Takers in regard to a career in ICT, thereby providing for the reader 'reality backdrop' by which to appreciate the strength of the remarkable attitudes expressed by Non Taker respondents to the high school survey. Figures 4–6 demonstrate responses of the women to some of these stereotypes.

Figure 4. Responses of women working in ICT industries (n = 272) to a question regarding the use of imagination and creativity in their career

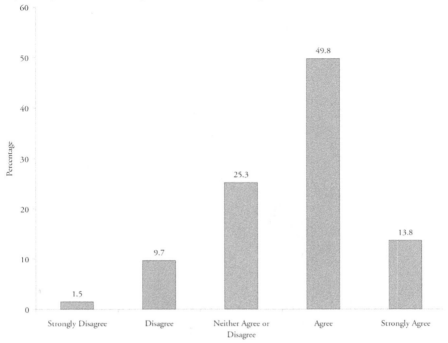

ICT stimulates imagination and creativity

The frequencies displayed in Figure 4 show clearly that from the perspective of most (64%) respondents to the Women in ICT Survey, their career is one where they have the opportunity to use their imagination and creativity. This provides a striking counterpoint to the finding (which will be expanded in the subsequent section) of a prevalent perception among Non Taker respondents to the high school survey that ICT careers do not provide opportunities for using imagination and creativity.

Figure 5 demonstrates that 62% of respondents to the Women in ICT Industry Survey disagreed that a career in ICT is sedentary. This reality for women working in an ICT career contrasts strongly with perceptions of respondents to the Girls in ICT Survey (Non Takers) whose qualitative comments tended to suggest perceptions to the contrary. It is noted, however, that 20.3% of respondents did agree with this perception, it is also noted that 8.1% of respondents to the Women in ICT Industry survey agreed that "ICT is boring" and a further 11.9% neither agreed or disagreed, indicating some ambivalence towards this particular item. Discussion of some difficulties encountered by women respondents to the survey may be found in Anderson, Timms and Courtney (2006), Courtney et al. (in press) and Timms et al. (in press);

however, exploration of this particular topic is beyond the scope of the present chapter.

Figure 5. Responses of women working in ICT industries (n=272) to a question regarding sitting in front of a computer all day

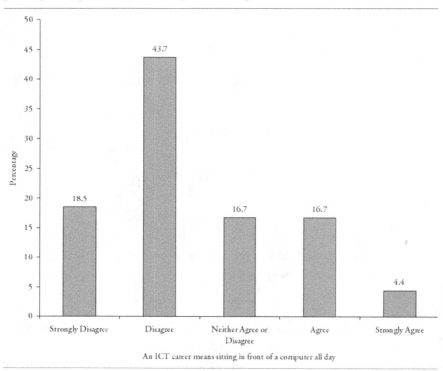

An ICT career means sitting in front of a computer all day

Of the respondents to the Women in ICT Industry Survey, 75.4% tended to agree that a career in ICT enabled them to work with others, thereby providing a contrast to a persistent perception of Non Taker respondents to the Girls and ICT Survey that a career in ICT was destined to be solitary.

The findings from the Women in ICT Industry Survey have been presented at this juncture within the current chapter to provide a background of reality to the findings from the Girls and ICT survey and focus groups which will be provided next. It was of interest to researchers that women working within the industry tended to disagree that ICT was boring, lacked opportunities for the use of imagination and creativity, did not provide opportunities to work with others and involved sitting at a computer all day. As will be seen in the following section, these perceptions featured highly among school girl respondents (Non Takers) to the Girls and ICT Survey, in regard to their decision not to undertake ICT at an advanced level in senior high school.

Figure 6. Responses of women working in ICT industries (n=272) to a question regarding the social nature of an ICT career

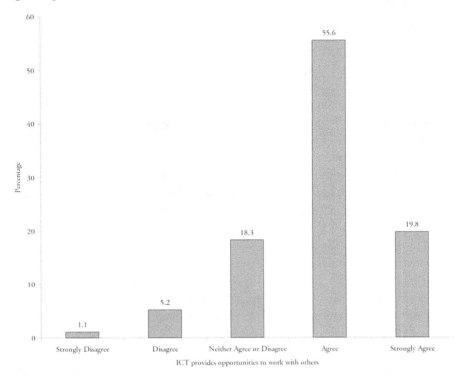

ICT provides opportunities to work with others

Emergent Themes from Girls and ICT Survey Findings and Comments

The Girls and ICT high school survey allowed space for respondents to provide comments, these provided researchers with the opportunity to explore the quantitative findings at greater depth. Analysis of these data from the Non Takers (n = 1322) uncovered three major themes which shed light on the quantitative findings:

»   Non interest because computers are perceived as boring
»   Computers are not perceived as central to the students' future career
»   The subjects were perceived as too hard or too technical for serious consideration as an option for girls to study at high school
»   On the other hand, comments from the 130 Takers tended to fall into two themes:
»   Intrinsic interest
»   Computers are central to future career aspirations

These themes have a polar relationship with the first two themes articulated by Non Takers. A polar opposite of the third theme identified by the Non Takers, the subjects perceived as 'too hard or too technical,' was not identified in the responses of the Takers.

Non Takers

Theme 1. Non-interest because computers are perceived as boring.

The first theme accounted for the largest proportion of comments by Non Takers with the majority of respondents commenting along the lines of "I am not really interested in using computers," or "it's boring," without any elaboration. Of particular interest to the research team were the more complex comments which offered more insight into the thought processes of participants. Among these perceptions was an understanding that computing subjects led to solitary careers and alienated them from the rest of society "I would rather interact with people than a computer." Recent research (Victoria State Government, 2001; Young, 2003) had similar findings in regard to this perception, stating that the majority of young people have a predominantly negative perception of ICT careers as being solitary, sedentary and consisting of low level tasks, interacting only with a computer rather than people. One participant expressed this by writing "computers are too difficult to understand and I'd rather not be stuck in a room typing all day." In another example, a Non Taker endorsed this particular stereotype by asking, "have you ever thought that computers and sitting in front of one all day is boring and draining?" Girls perceived that the use of computers was synonymous with a lack of social interaction. This was a common premise in the review of the literature conducted by Adya and Kaiser (2005). Therefore, it would seem that the attitude of girls towards a career in computing is based on a pervasive and persistent stereotype.

The AAUW (2000) reported that girls tended to dislike the technical and narrow focus which characterises advanced computing subjects within high schools. Participant comments in the current research are supportive of this view: "It's boring in high school. I find computers incredibly interesting and would love to have studied ICT in high school and continued that study through to university if it was more interesting." Others reported the subjects were lacking in practical application "because I personally think that sitting in front of a computer is boring and I am more of a 'hands on' person." The AAUW research also suggested that girls saw computers as instrumental to tasks "useful primarily for what it can do" (p. 9) rather than inherently interesting in their own right. This sentiment was summarised by a participant who commented, "I am not interested in what happens with my computer, all I want to do is use it." Other comments indicated a perception that extra time spent with computers would be dangerous to one's health, "I would not want to be stuck in front of a radiation screen!" and "Computers give me headaches." Another participant wrote "I'm not interested in that field of study, sitting in front of a computer screen makes me sleepy." These last comments are perhaps an indication of student disengagement and boredom referred to by the Victoria State Government (2001) and by Margolis and Fisher (2003).

The Victoria State Government (2001) recommendations included the comment "Young people are particularly interested in knowing what is actually involved in different jobs—what people actually do in their average day. They are interested in material that helps them 'visualise' what a job would be like" (p. 6). It is clear from respondent data reported in this survey that girls think that technology related careers involve the same activities they participate in at school level, and are consequently ill informed of the diversity of interesting and challenging career options available within the ICT sector.

Theme 2. Computers are not perceived as central to the students' future career.

Of particular interest to researchers was the perception that computing careers were not creative. One participant stated, "I'm into more creative things," and another participant expanded this into her future career path, "I wish to follow a more creative career path that involves me moving around and thinking more." Yet another participant was more specific in her assessment of 'creative careers' in comparison to her perception of advanced computing subjects, "I like subjects such as art and home economics and I'm not really into ICT because I don't think it's very creative." In addition, among student responses a stereotypical view of computer scientists emerged "I don't enjoy it, not my type of subject. Not interested. Geeky," and further, "Not interested/boring/nerdy." These perceptions support Frieze's (2005) observation that the predominant societal view of ICT is that the field is "populated by geeky guys, while at the same time the image of the field itself is seen as little more than coding" (p. 397). They are also consistent with Frieze's reflection that "The image of computer science as a broad and exciting field with the potential for diverse participants is, for the most part missing from the big picture" (2005, p. 397) and the Victoria State Government's (2001) report that awareness of how technology is shaping the modern workplace is largely missing from information which is provided to students within the school system.

Girls appeared to be generally poorly informed as to the reality that skills previously regarded as the preserve of ICT specialists are now expected of workers in most organisational and occupational milieus (Victoria State Government, 2001). For example, students interested in career pathways such as science, medicine, hospitality, military and agriculture commented similarly that computer skills were not relevant to these careers. Two particularly striking comments were, "I am wanting to join the RAAF as an Intelligence Officer" and "It wasn't needed for my chosen career path in science/medical studies." In addition, a number of participants indicated that their present level of skill was adequate, "It just doesn't interest me. I know the basics about computers and that'll do me" and "we had done basic computer skills in year 8/9 so I didn't feel the need to learn anything more than the 'basics.'" Finally, some participants expressed views commensurate with denial of modern technological society, for example, "I'm not interested in a world dictated by technology."

Theme 3. The subjects were perceived as too hard or too technical for serious consideration as an option for girls to study at high school.

The girls' comments reflected perceptions that the subjects were particularly hard and dominated by intensely difficult theory. Examples of this perspective included, "[I'm] not interested in algorithms and studying computers"; "It [IPT] looks too complicated, hard and boring"; and "I have heard they are difficult. I believe they (the classes) get too technical, which scares people/girls away from the class." This was consistent with the findings of the AAUW (2000), which suggested that advanced computing subjects in high school and introductory computing subjects at the college level are used as a 'filter' rather than a 'pump' and are designed to "filter out students rather than invite them into the discipline" (p. 47). It would appear that the hydrological analogy is enduring and the connection between this and Camp's (1997) 'incredible shrinking pipeline' is obvious. It is further noted by the AAUW that this use of the subjects may serve to alienate many students who might otherwise have excelled in computer science, but who lack the necessary experience to achieve in the subjects at this level. Participants did refer to their perception that prerequisite knowledge was necessary to do well in IPT/ITS and that consequently it was not worth persevering in the subject, "I have not any ICT lessons before and feel I would lag behind in senior classes"; and further that, "Because I did not do the prerequisites. These are not needed in the future."

Among respondent comments testifying to perceived complexity and difficulty of computer theory were references to Queensland's Overall Position (OP) scoring system. This system is used in the tertiary selection process by the Queensland Tertiary Admissions Centre (QTAC, 2006); however, individual universities set their own "cut offs" for admission into particular courses. Consequently in Australian states, a senior secondary student's academic study is dominated by a ranking system which will determine not only the career pathway they will be in a position to enter, but also the university they will be able to attend. A recent publication from the state education authority, Education Queensland (2006), advised parents, "students should work to their strengths and select subjects they are best at, enjoy and suit their future studies" (p. 13). It therefore follows that year 10 students, who are on the cusp of choosing their senior high school subjects, are advised by teachers, parents and counsellors to choose subjects that will not compromise their chances of achieving the best possible OP score. The spirit of this advice is acutely reflected in a number of responses from participants in the survey, "I chose other subjects that I'm better at, to improve my chances of a good OP" and a further perception that advanced computing subjects "would make my OP go down because I'm not smart enough." In addition, the anxiety associated with the pressure to obtain the best possible OP was reflected, "I don't think I would be good at it. Don't want to screw up my OP any more."

Consequently, student familiarity with subjects in the junior school provides a basis for their subject choice when they enter year 11. Several participants indicated that they were "put off" by their experience of computing subjects in junior years "CTY (Computer Technology) in years 9 and 10 were boring and the content was very primitive"; "I did it in Year 9 and 10 and found it boring"; "I studied ICT in my junior years of high school, and strongly regret it, as I found I hated this subject." Yet another participant indicated that some of the responsibility for her dislike of previous computing studies was due to lack of assistance in the learning program, "I found computer studies in previous years difficult and I did not really get the help I needed to improve this problem." Another respondent referred to a possible lack of ICT support within her school, "Computers are crap and they are boring and I don't like working on school computers cause they are crap and never work properly."

Hence, teacher preparation and professional development, curriculum issues and maintenance issues featured within survey comments from Non Takers. These issues have appeared before in the literature. A Ministerial Council on Education, Employment, Training and Youth Affairs (MCEETYA) (2003) report indicated that many teachers may well be teaching outside their area of expertise, thereby creating an area of anxiety for the teachers involved and frustration for students. The AAUW (2000) emphasized the importance of teacher expertise in conveying enthusiasm and positive attitudes to students. In addition, Granger, Morbey, Lotherington, Owston and Wideman (2002) commented on the need for well maintained equipment and teacher education in ICT use:

Teachers can neither teach computer skills nor integrate ICT into curricula without having at their disposal computers that work. Clearly, a lack of appropriate material resources inhibits learning and causes frustration and resistance in school communities. Further, appropriate full-time technical support and significant opportunities for teacher education in ICT use are as necessary as up-to-date equipment if teachers are to move toward curricular integration and meaning-making (p. 487).

Turner, Bernt and Pecora (2002), who surveyed a number of women employed in ICT careers, found that many very successful women working in ICT cited their school experience of computing as their first introduction to the field. Furthermore, they took discouragement or encouragement from teachers in high school very seriously. A number of girls indicated that they were influenced by teachers, for example,

"In junior the teacher made me dislike it and there wasn't much 'fun' involved." Others indicated discouragement at subject selection evenings, "When I went to the subject selection evening the subject had seemed hard and required a lot of time and effort."

A number of students indicated that they did not feel that they were capable of achieving in ICT on the basis of information that they had been given, "It is put across that only the smarter people (usually boys) do this subject" and others that they were actively discouraged, for example, "Certain Counsellors [sic] made it seem dull." This last comment supports the observation of Jepson and Perl (2002), "school guidance counselors don't have sufficient knowledge to distinguish between computer use and computer science—because guidance counselors don't know, the students don't get to know" (p. 37). It was also found by von Hellens and Nielsen (2001) that careers in ICT were not promoted well in high schools in comparison to those in other technical areas and warned, "a school's influence is crucial" (p. 51).

Takers

Theme 1. Intrinsic interest.

According to the Victoria State Government (2001), today's youth are the 'show me' generation (p. 8); they will consider interesting options so long as they are convinced that these claims are true. Of the Takers, some comments regarding intrinsic interest ranged from "The subject is fun!" to "During school studies in ITS and IPT I found game development and programming interesting and fun, so I wish to further my skills in these areas"; and "ICT is an interesting field of study. It incorporates maths and logic and I enjoy the subject." It is anticipated that analysis of focus group sessions will shed further light on influences which have propelled Takers to a viewpoint where computers are seen as an exciting and positive field of study.

Theme 2. Computers are central to future career aspirations.

Unlike the Non Takers, who indicated their perceptions that advanced computing subjects would not be relevant to their futures, Takers generally supported Margolis and Fisher's (2003) observation that computing science offered versatility because almost any field is computer-related. Some comments included, "Plenty of careers available in this field"; "Computers are used a lot these days and I thought by doing this subject I would gain useful knowledge of computer systems"; and "It's common usage in all areas of industry/work. It has become an 'essential' tool of everyday learning/leisure/work, etc." Other comments were, "Plenty of careers available in this field"; "I think that even if I don't study this subject at uni, it will help me in the future"; "Growing industry. I enjoy it a lot and it's good money"; and simply, "It's easy and I'm quite good at it."

Focus groups

Eight focus groups were conducted: two in Cairns (a large regional city in Far North Queensland) at a large state high school; two in Atherton (a rural town in Far North Queensland) at a state high school; two in Brisbane (the largest city in Queensland located in the South East corner of the state) at a large state high school, and two in Toowoomba (a regional centre in South East Queensland) at a non government girls' school. A total of 48 grade 11 and 12 girls contributed to focus group discussion. It was planned to interview girls who had participated in the 2005 survey, and indeed most of the participants had been survey respondents some had not; but they were included because they were keen to contribute to the research. Researchers also faced the reality that the end of school year had intervened between the two stages of research and all of the girls who had been in grade 12 in 2005 had left high school.

Focus group findings.

As most focus group participants were Takers of advanced level ICT (n = 31, 76%), they were keen to contribute to understandings of ICT subjects. They regarded advanced ICT subjects highly, hence they were unable to extend researchers' understanding of the second theme (computers were not seen as relevant to the students' future careers). This was because focus group participants were generally cognizant of the potential role of technology in the workforce and society of the future:

It will probably help with our future as well because of how much technology is advancing. Most likely most workforces will have some aspect of computers in it, so even though it is more than common knowledge of computers, it is still good to have that knowledge anyway.

However, participants were able to open a window for researchers into the world of the school student, thus providing powerful and mature insights into why computing subjects were seen as boring and the subjects were perceived as too hard or too technical for serious consideration as an option for girls to study at high school by so many survey respondents. The finding that Non Taker respondents to the high school survey considered advanced computing classes boring was shared with focus group participants by research team members. Takers and Non Takers of ICT were asked where they thought the mind-set came from. While Takers and Non Takers differ markedly on their attitudes to advanced ICT classes, they were remarkably similar in regard to the source of Non Takers' perceptions that ICT was boring.

Theme 1. Focus group commentary on survey finding that Non Takers considered computers boring.

The finding that Non Takers in the survey considered advanced computing classes boring (Timms et al., 2006) was shared with focus group participants by research team members. According to participants, the perception began with the experience of junior high school ICT subjects as highly structured, with uninspiring and boring tasks which had no apparent purpose. An example of this was, "they don't really tell you how to connect. I know they are trying to break it down but I sometimes feel like I can't apply that stuff to this stuff." There were also references to the prescribed nature of junior high school computing subjects "when you are in Junior— well for me they made ICT really quite boring. Like the subjects were planned and there was no leeway for anything. You kind of had to do this and this." This would suggest a lived experience of a series of disconnected and apparently purposeless tasks, specifically at the level of junior high school.

The expressed need for contextual activities is consistent with observations within the literature that girls tend to be more interested in using the computer to accomplish a particular task (Lynn et al., 2003). This is also consistent with the research of the AAUW, who found "girls are concerned about the passivity of their interactions with the computer as a tool" (p. ix). Unfortunately it would appear that concerns expressed by Frenkel in 1990, "Computer science curricula place an emphasis on step-by-step division of functions and women lose interest" (p. 1) and have failed to diffuse into the understanding of those who plan and prepare curricula within the junior high school.

A number of focus group participants mentioned that their early experience of computing subject was largely comprised of learning keyboard skills:

It is probably a really hard way to start it. They start with typing that is really bad. I can see why you need to learn to type but it is not a very good introduction. A lot of people don't get past that.

Of particular concern to the research team was a perception, which is apparently widespread, that the content of junior high school subjects was reflective of the more specialized subjects in senior high school, one participant mentioned, "Grade 8 Computer subject, the course is really boring, just typing. I know so many people who said [advanced computing] is boring, because you just do typing." Another participant in another school indicated her shared experience of a less than inspiring start to her awareness of an advanced computing subject: "for me it was really boring and I think most people are expecting Senior to be exactly the same, but it is not."

Students referred to the difficulties encountered with teacher expertise, which is apparent even within the advanced senior computing subjects:

Yeah, too hard. When you get a bug in the system and the teachers just say "Oh, I don't know," and then your whole assignment goes down the drain because you can't work out this bug and then the teacher says, "Hey, hold on, it is just a mystery," that really puts you off. Especially if you have to start again. It is like you got to a certain point and there is just this little problem and it is like that is too hard.

It is posited that if this particular problem arises in senior advanced level computing classes, how much more likely it is to occur at a junior level where, because all students take the subject, there are more classes. For that reason, it follows that many teachers of the junior subjects may well be teaching outside their area of expertise. In fact in 2003 a Ministerial Council on Education, Employment and Youth Affairs (MCEETYA) report 'Supply and demand of primary and secondary teachers in Australia' noted:

> It has been suggested that some teachers working in these subject areas [mathematics, sciences and ICT] are often not well qualified to teach in their subjects and are working out of the subject area for which their qualifications cater (MCEETYA Part F, 2003, p. 24).

Several focus group participants made comments that indicated that their own knowledge of computing exceeded that of the teacher taking junior high school classes: "Sometimes the teachers don't know what they are teaching. That is when you have to really work it out by yourself. I remember a class where we were doing programming and the teacher had no idea."

Conversely, but reinforcing the importance of teacher expertise, other focus group participants indicated that they had been encouraged by positive experiences in junior high school, an example is found in this piece of focus group dialogue in response to a researcher question as to people who may have been instrumental in encouraging students to take up ICT at a senior high school level:

> GIRL ONE: Miss Y, she has already left but she is actually sort of our driving force. She got me into computers.
>
> RESEARCHER: And how did she inspire you or stir your interest up?
>
> GIRL ONE: She knew so much. And she could show you everything, everything you wanted to know really.
>
> GIRL TWO: We had just met her and she watched us work and then introduced us to designing and stuff and then she starting using us to make things. She was good. She encouraged us.

The teacher referred to by these participants demonstrated familiarity and fluency with technology, which the students found inspiring. This sort of fluency was referred to by the AAUW (2000) in their recommendation to education authorities to,

> Prepare tech-savvy teachers. Schools of education have a special responsibility: They need to develop teachers who are able to design curricula that incorporate technology in a way that is inclusive of all students. Schools of education also must be able to assess "success" for students and teachers in a tech-rich classroom. The focus for professional development needs to shift from mastery of the hardware to the design of classroom materials, curricula, and teaching styles that complement computer technology. (p. xii)

The previous focus group comment illustrated two other prominent themes within the literature in regard to encouraging girls to undertake advanced computing: the teacher concerned provided girls with a positive role model of female technological expertise (McNair et al., 2001; von Hellens et al., 2001); and girls' need for encouragement and support (AAUW, 2000; Turner et al., 2002).

Theme 2. Advanced computing and future careers.

As previously mentioned, focus group participants were generally not enlightening in regard to the second theme to come from Non Taker survey respondents that ICT was not seen as relevant to their future careers. This was because they tended to recognise the value of ICT and were generally concerned about the "bad press" that surrounded computing and computing subjects. Relevant comments included, "you get to choose all your subjects, so why not choose one that is going to benefit you later in life," "nowadays everything is going computer wise so you would be silly not to do it" and, "I thought it looked interesting. It had lots of opportunities to do stuff later in life."

Theme 3. Advanced computing is too hard or too technical.

The third theme to come out of the Girls and ICT Non Takers' survey comments was that the subjects were perceived as too hard or too technical for serious consideration as an option for girls to study at high school—this theme was also addressed in focus group discussion.

Non Takers who responded to the survey indicated that they were uninformed as to what senior ICT subjects could offer them and the career opportunities that awaited them—in the absence of this information they chose not to do ICT, when it became apparent to them that subject choice was crucial to their ability to enter the university and course of their choice. Perhaps the most enlightening comment on this point comes from a particularly articulate Non Taker who participated in one of the focus groups—she explained that the interaction between lack of knowledge of the skill set offered by senior ICT subjects and the imperative to perform well at a senior level for the purposes of achieving useful tertiary entrance scores created a situation where computing was not even considered as a possible career pathway:

...the computer study thing, people in junior school don't give you any indication of what it is going to be like. They don't do any of the basic work leading up to that, those kinds of subjects. Like Maths C, it is like a whole new area of work. You go into senior school, and people do take their senior studies seriously when they are choosing what to do. At least with something like Chemistry you would have a basic idea from junior science what kind of field you are going into, but with ICT it is like...an unknown entity.... People are going to choose it and, the way the subject change is now, you have got to stick it out for two terms really before you can change it and people go, "Well, what the heck, give it a go" and you really despise it and you have to stick out with it and still that might count for your OP,...a lot of people aren't willing to take that risk.

This comment is consistent with previous observations of De Palma (2005; 2006) that the body of knowledge that constitutes 'computing' is not as easily defined as those bodies of knowledge that constitute core material in subjects like science (mentioned by the last participant). De Palma's opinion was that computing knowledge (and possibly awareness that one has a propensity for computing) is acquired by informal means and trial and error (the 'tinker factor'). This view was supported by focus group participants and possibly warrants further investigation.

Conclusion

The current research project has highlighted a number of areas that are worthy of further investigation. These are:

» There is a significant difference in attitudes between Takers and Non Takers of advanced computing subjects as to the interest and relevance of ICT subjects.

» There was little support for previous research findings that negative attitudes of girls to ICT subjects are based on abhorrence of male emphasis in computer gaming or intimidation of girls in male dominated classrooms.

» Girls were 'put off' ICT by boredom in computing subjects at the junior high school level. This boredom translates into enduring and inaccurate stereotypes about advanced computing subjects and ICT careers.

» The imperative to perform highly at tertiary entrance level does factor into student decisions not to undertake advanced computing. This decision is often based on a negative experience in the junior high school.

These may well be crucial to an understanding of how girls can be empowered and become confident to enter the ICT field. It is noted that as far as ICT is concerned, girls make choices as to their senior high school subjects and therefore embark on a career pathway at the end of grade 10 with very little knowledge as to what is involved in a career in ICT. Focus group participants emphasised that their experience of subjects at junior high school level was pivotal in their choice of subjects for senior high school. This then leads to questions of skills for the teaching of junior computing subjects—maybe in the form of intensive professional development for existing teachers and university level preparation for pre-service teachers or some other measures not yet determined such as access to user friendly web modules that provide help with skill development and content. This point is consistent with recommendations of the AAUW (2000), and the Victoria State Government (2001) report:

There was almost universal concern among all stakeholders that IT teachers were not adequately trained and resourced. There was also a high level of concern from industry that schools were not sufficiently emphasizing the broad application of IT skills in the modern workplace, and the need for IT graduates to also have strong commercial knowledge, skills and instincts. (p. 43)

Teacher expertise in ICT at the junior high school level is a key component of the sustainability of advanced computing subjects. Focus group participants stressed that their junior high (middle school) experience forms a basis for their subject choices in the senior high school. Turner and colleagues (2002), who surveyed a number of women employed in ICT careers, found that many very successful women working in ICT cited their school experience of computing as their first introduction to the field. Furthermore, they took discouragement or encouragement from teachers in high school very seriously. It was also noted by von Hellens and Nielsen (2001) that careers in ICT were not promoted well in high schools in comparison to those in other technical areas and warned, "a school's influence is crucial" (p. 51). Comments from focus group participants would appear to support von Hellens and Neilsen.

The current research has confirmed a significant difference between the attitudes of Takers and Non Takers of advanced computing subjects towards those subjects, with Takers expressing much more favourable attitudes than those expressed by Non Takers. Many of the usual reasons provided within the literature for the difference in attitudes and for the ever reducing numbers of women entering courses which lead to careers in ICT, while logical and persuasive, were not supported by survey or focus group participants or did not feature as prominent reasons for the poor take-up of advanced computing subjects. Consistent with previous findings of the Victoria State Government (2001), qualitative data from the survey comments and from focus group participants indicated that students within schools who do not take advanced computing subjects were generally poorly informed as to how technology is reshaping the modern workplace and that they will need to develop skills in ICT. This lack of knowledge combined with the imperative to perform as well as possible in order to achieve the best possible tertiary entrance score have to be considered as possible factors in student decisions not to consider ICT as a career pathway.

Because choice of subjects at the end of junior high school is central to career pathways students eventually undertake, the current research therefore throws a spotlight onto computing at the junior high school level. This would indicate that the role of the school in providing girls with affirmative experiences of technologically conversant teachers (AAUW, 2000), well maintained equipment (Granger et al., 2002) and state of the art information about the importance of technology in career pathways (Victoria State Government, 2001) are perhaps the most fundamental contributors to 'pumping the pipeline' and providing students with the ability to make more informed choices about their career pathways.

. . .

PART 2

Intellectual Disability and ICT

Building of Competence Case Study:
The Initial Situation

Linda slowly turned the pages of her book, staring intently at the illustrations. Around her, the other students, all approximately nine years old and in their fourth year of schooling, also concentrated on reading. Each student read from a book deemed to be at an appropriate reading level and belonging to a classroom series designed to interest students of this general age group. The classroom teacher and teacher's assistant circulated around the room listening to individual students reading aloud while the others continued with silent reading. Eventually, during the reading session, all the students in the class would have a turn to read aloud to one of the adults. When it came to Linda's turn—something different happened. Linda did not read the actual words but made up a story based on the illustrations in her book. This would not be accepted from any other member of the class, but Linda had not managed to learn basic reading skills during her first three years of schooling in a segregated special education unit for students with intellectual disabilities or during her first year of inclusion in a regular, general classroom. Linda did not create a fuss if she was allowed to make an interpretation of the story based on illustrations but became angry, uncooperative and disruptive if any attempt was made to teach decoding skills or anything that meant seriously engaging with text.

When it was time for a class or individual writing task, Linda mimicked the other students by writing some words and letters on the page but did not attempt to write anything that made sense. This was similar to the tactic used by Linda for reading in that she could appear to be doing the same work as her peers and would not be pressured to try to make sense of the letters and words and would be largely left alone if she didn't draw attention to herself. She had given up on the idea of extending her reading or writing skills and had developed a very effective defense against any attempts to engage her in reading or writing instruction. Linda had previously participated in individualized remedial sessions at the special unit for students with disabilities since beginning her enrolment at the regular school but had now totally given up on making any progress and simply did not want to have anything to do with reading or writing unless it was unavoidable. Linda was physically included in the classroom, and on the surface she was undertaking the same or similar tasks as her general education peers, but she wasn't actually demonstrating any academic competence at the same or similar level of her peers and was not really socially accepted by her classmates. Perhaps the one activity that Linda thoroughly enjoyed was using one of the classroom computers.

Inclusion of students with disabilities: Background to the case study

An escalating number of students with intellectual disabilities are being included in regular classrooms. In the United States, the Individuals with Disabilities Education Act (IDEA, 1999) and the subsequent reauthorized Individuals with Disabilities Improvement Act (IDEIA, 2004) requires that students are educated in a natural and 'least restrictive' educational environment, wherever possible with their non-disabled peers. The Act states that "to the maximum extent appropriate, children with disabilities, including children in public or private institutions or other care facilities, are educated with children who are not disabled and special classes, separate schooling, or other removal of children with disabilities from the general educational environment occurs only when the nature or the severity of the disability is such that the child cannot achieve academically in general education classes with the use of supplementary aides and services" (Etscheidt, 2006, p. 167).

This legislation is mirrored in many other countries and also in the 2006 United Nations Convention on the Rights of Persons with Disabilities. This convention emphasizes the right of people with disabilities to access an inclusive education and flows through to international law. Inclusion of students such as Linda therefore should be commonplace, but Smith (2007) reported that little progress has been made in the United States to include students with intellectual disabilities in regular classrooms. Ferguson (2008) noted that in 1989, 6.8% of students with intellectual disabilities were included and, by 2004, only 13.1% of these students were included in regular classrooms. Ferguson also points out that much progress has been made with the inclusion of children with other types of disabilities. In Australia, according to Grace, Llewellyn, Wedgwood, Fenech and McConnell (2008), the small number of successful inclusions evident in their study occurred in spite of the current government policy rather than because of it. It appears that although inclusion has been widely accepted for students with disabilities, the students with intellectual disabilities have experienced a more modest level of inclusion in regular schools. Nevertheless, it is a reasonable assumption that the number of students with intellectual disabilities in regular schools will continue to increase, making the scenario outlined in the first paragraph more common.

The other feature outlined in the brief vignette concerns the quality of the inclusion. If a school environment is truly inclusive, it will involve more than mere physical inclusion. In Linda's fourth year of schooling, she was merely mimicking the behavior of her classroom peers but making very little academic progress. Her lack of academic competence also had a negative effect on her social competence and acceptance by peers. Simply placing students with intellectual disabilities in general education classrooms without strategies to improve academic competence virtually guarantees difficulties with peer acceptance. Academic and social failure of students with intellectual disabilities in regular classrooms then adds fuel to the ongoing debate about appropriateness of placement. Despite the clear mandate of current legislation, O'Rourke and Houghton (2006) recently highlighted the ongoing debate amongst educators about whether students with intellectual disabilities gain academically or socially from placement in regular classrooms and the polarization of opinions.

They also point out that many students with intellectual disabilities find regular classrooms to be intimidating if support structures and appropriate programs are not in place. Linda's negative reaction to reading or writing instruction appears to be quite prevalent with this particular group of students. For example, Martin, Martin and Carvalho (2008) argue that children with disabilities often experience psychological and emotional reactions when dealing with reading failure and that this is frequently results in pain and the development of a destructive and negative mindset towards reading. They report that many students with intellectual disabilities feel shame, frustration and embarrassment at their lack of ability to read and function optimally in the world in the way that their classroom peers can. A negative mind-set was very apparent in Linda's behavior, and this aversion to text proved to be too difficult an impediment for her teaching team to overcome in her fourth year of schooling. In the words of Filler and Xu (2007, p. 94) "participation and not mere geographical proximity is the necessary pre-condition for achievement, and so meaningful participation requires systematic planning."

Planning and Implementing an Inclusive Program for Linda

At the beginning of Linda's fifth year of schooling it was decided to experiment with an individualized program that took advantage of her positive attitude towards computers. It seemed logical to try to introduce elements of the regular curriculum such as computer skill development, reading and writing development and studies of society via an information communication technology enhanced approach. While Linda resisted attempts to encourage or coerce her to participate in traditional pedagogical approaches, she seemed very content to be working on the computer. Some risk was involved in centering her work around computing since it was not known if Linda would exhibit the same aversive pattern of behavior if the computer based work became more challenging or involved tasks that engaged her with text. Since the program was to be individualized, adjustments to the difficulty level could be made as the program progressed. A literature review by Filler and Xu (2007) revealed that many researchers recognize the importance of instructional flexibility and recommend individualized approaches for students with intellectual disabilities. Other authors such as O'Rourke and Houghton (2006) point out the pitfalls and risks associated with individualized programs for students with special needs. It was recognized that Linda's very negative attitude towards classroom work associated with reading and writing was one of the major impediments to the development of competency; therefore, the individualized program needed to be accepted positively by her.

While many teachers would approve of individualized programs that are 'easier' than the work normally completed by the regular class members, O'Rourke and Houghton cite studies that confirm that students with intellectual disabilities detest individualized work that is obviously at a lower level than given to the rest of the class. They feel that this results in being mocked by other class members or that their peers would have a very low opinion of them. This may explain Linda's preference to mimic the same work that the other students were doing rather than participate in individualized remedial reading or writing classes. At the same time, it would be foolish to plan work that would lead to continual failure and frustration due to inappropriate difficulty levels. In most countries, including the United States, Australia and Canada, the work program and adaptations planned for students with intellectual disabilities is recorded in the IEP (Individualized Education Plan). According to Jorgensen, McSheehan and Sonnenmeier (2007) the judgments that teachers make about students' competence have a profound effect on the specific features of the IEP and the extent that the child is included. In this case a gamble was taken that Linda could develop some higher order ICT competence and that she could develop competence in reading and writing along with developing some subject specific competence. Jorgensen (2007, p. 251), building on research by Donnellan, proposed that "the least dangerous assumption is to presume a student is competent to learn general education curriculum and to design educational programs and supports based on that assumption." Assuming incompetence would mean that the student would not be afforded opportunities and would be more harmful.

The program in a nutshell

Now returning to the intervention design: the basic kernel of the program was to teach Linda to use desktop publishing software in a way that initially avoided text, so that she could master some of the complexities of the software without being anxious about text construction. It was based on the student producing a variety of documents related to her chosen theme of jungle animals. Various templates and models of document types were introduced to the student over time. Avoiding the use of text in the first phase was achieved by concentrating on other elements of the software such as inserting borders, graphics and photos. This also provided a link to the regular curriculum since she was very interested in animals, and at the time, the class was participating in a unit about how animals adapt to the environment. Another skill to be taught was how to use a search engine on the Internet to locate and save relevant pictures. It was planned to use the pictures to stimulate discussion about animals and to supplement the graphics and information found on the web with books. Gradually the use of text was to be introduced as Linda became more comfortable with the program. As she progressed, different topics would need to be chosen since an important element of the program involved choice. In the introductory period Linda decided to concentrate on jungle animals in particular since she was more interested in this category. Later in the program peer tutoring became an important element, but this occurred naturally and was encouraged, rather than planned for at the outset. Desktop publishing (DTP) was deliberately chosen because the other students knew how to use word processing software, but DTP was considered to be more complex and unfamiliar. As Linda's competence grew, she discovered that other students wanted to be shown how the software worked and also developed a respect for Linda's competence in an area that they weren't confident or knowledgeable.

An important element of the program involved avoiding past negative mind-sets by tapping into an area of interest, which was working with ICT. It aimed to offer as much choice as possible, gradually introducing text for building reading and writing competence and building computer skill competence along with developing subject specific knowledge. Later in the implementation of the program, peer tutoring became very important. This occurred when Linda became the peer tutor for regular class members and was used to develop competency in communication skills. Another important strategy to be employed was encouraging Linda to talk about the steps she used to solve problems, thereby making the metacognitive processes explicit. The following section will examine the important elements of choice, the importance of developing competence for social acceptance by peers, ICT skill development, reading and writing development, peer tutoring, the development of oral communication skills, the development of metacognitive skills and the building of self-perception, particularly in the domain of academic self-perception.

The importance of providing choice

Providing as much choice as possible in selecting topics for embedding instruction and problem-based learning was considered to be a central element in the program. By involving the student in selecting areas of study, it was hoped that negativity could be eliminated or reduced. Stenhoff, Davey and Lignugaris/Kraft (2008) investigated the use of choice on task completion and accuracy in a group of students with learning disabilities and reported that earlier studies had provided evidence that offering choice to students with disabilities had led to stronger task completion and a reduction in behavior problems. A meta-analysis of the available literature led Morgan (2006) to conclude that offering choice led to better task completion, task engagement and accuracy. Earlier research had also demonstrated that choice reduced inappropriate behaviors by students with mild disabilities (Powell & Nelson, 1997). The recent research conducted by Stenhoff, Davey and Lignugaris/Kraft (2008) concluded that offering choice led to increased academic achievement. Differences in academic achievement that they noted included a high level of task completion and a better level of correctness/accuracy.

Building academic competence and the link to social acceptance

This intervention focused on building up competence in basic computer skills, reading and writing skills and subject specific knowledge. It also included an element of training in metacognitive processes. Siperstein, Parker, Bardon and Widaman's (2007) review of the literature revealed that advocates of inclusion believed that one of the main benefits of inclusion was that contact with students with intellectual disabilities would, over time, lead to more positive attitudes to students with intellectual disabilities and subsequently greater social acceptance, along with greater opportunities for social interaction. They cite studies by Hastings, Sonuga-Barke and Remington (1993); Helmstetter et al. (1994); Manetti, Schneider and Siperstein (2001) and Nowicki and Sandieson (2002) to illustrate their point that simply including students with intellectual disabilities in regular classrooms would not automatically lead to positive attitudes. They contend that negative attitudes are likely to develop unless "the contact works in such a manner that it breaks down existing stereotypes rather reinforcing them" (p. 436). Wong (2006, p. 72) also concluded that "the proximity and presence of students with disabilities in the general classroom does not automatically bring about positive attitudes. On the whole, students with disabilities were less popular, and active facilitation and thoughtful intervention seems to hold the key to positive social acceptance in the general class-room." Smoot (2004) also reported lower levels of social acceptance of students with mild intellectual disability than other class members.

Siperstein et al., (2007) large study consisted of 5837 participants in 47 school districts in 68 schools in the United States. This was a randomly selected sample of public middle school students across a representative set of districts and schools. The survey included a perceived capabilities scale, an impact of inclusion scale and a behavioral intentions scale, an academic inclusion scale and a non-academic inclusion scale. The researchers used several types of traditional analysis along with structural equation modeling on the study variables. The researchers developed an a priori model of youth attitudes presented in Figure 7.

Figure 7. Siperstein et al. (2007) a priori model of youth attitudes

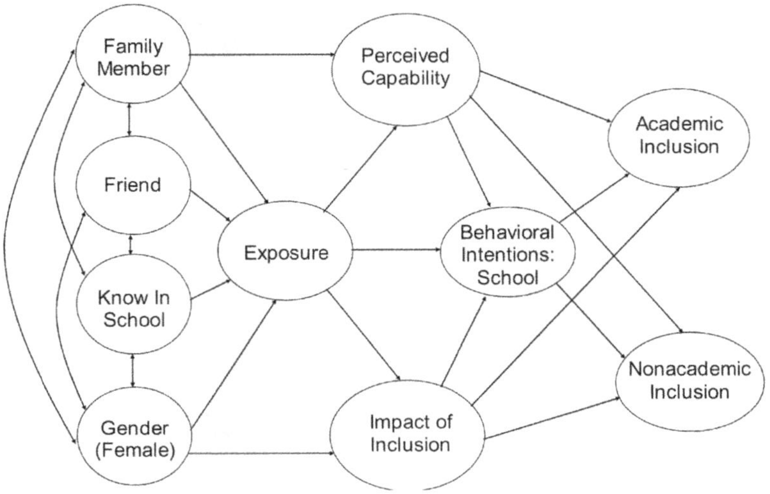

The main findings of the study were:

1. Contact between students with intellectual disabilities and regular students was lower than expected and still quite minimal.

2. Students perceived students with mild or moderate ID to have a more severe disability than the reality of the disability.
3. More students believed in the value of non-academic inclusion than academic inclusion.
4. Students believed that inclusion has positive and negative effects.
5. Perceptions of competence affects acceptance.

The most significant finding and the one most relevant to the following discussion on the use of ICT as a catalyst to positive inclusion was the power of perceived capabilities to influence academic and non-academic inclusion. Students generally believed that students with intellectual disabilities could carry out simple tasks but did not believe that they could carry out more complex tasks. Students' willingness to accept students with intellectual disabilities in academic and non-academic contexts was strongly influenced by how they perceived the capabilities of the students. Students were very concerned about the impact of including students with intellectual disabilities on their own academic performance. If they thought that this influence was positive, they would be much more likely to have positive attitudes towards the students with disabilities. A strong conclusion from the study was that "the findings from our structural equation modeling underscore the key role that youths' perceptions of the competence of students with ID play in their attitudes to inclusion, as well as their behavioral intentions to interact with students with ID in schools" (p. 451). The strength of the 'perceived capability' variable can be seen in the final model produced by Siperstein et al. (2007).

Figure 8. Siperstein et al. final model of youth attitudes

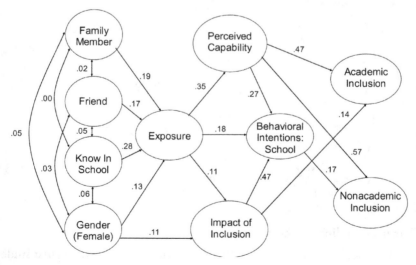

Earlier it was stated that an assumption was made that Linda would be able to develop competence in using the computer operating system and the associated software and that she would be able to achieve satisfactory results integrating ICT with the curriculum. Jorgensen, McSheehan and Sonnenmeier (2007) pointed out that little consensus had been reached amongst educators and policy makers about the appropriate achievement levels for students with intellectual disabilities. They also outlined how IQ based measures of competence along with tests of adaptive behavior were often used to make decisions about the overall competence of the students with intellectual disabilities. A warning was given about relying on standardized tests, and a summary of the different weaknesses of the instruments was presented. The American Association on Intellectual and Development Disabilities (AAIDD) argues that the relationship between competence and intellectual disability requires "a multidimensional and ecological approach that reflects the interaction of the individual with the environment, and the outcomes of that interaction with respect to independence, relationships, societal contributions, participation in school and community, and personal well being" (AAIDD, 2007). Linda's program was designed to maximize opportunities to develop basic competence but, more importantly, to boost her problem solving and higher order thinking skills, thereby building higher levels of competence that her peers would recognize.

Building ICT competence

Returning now to Linda's intervention, as discussed previously, desktop publishing was chosen as the key piece of software to facilitate and document Linda's work in the first unit associated with jungle animals. While desktop publishing tools operate in a different way to the familiar tools of word processing and were perceived by other students as being more challenging, they did not prove to be too difficult for Linda to master. The basic tools of desktop publishing allowed her to easily produce a variety of borders, insert photos and graphics and experiment with different document layouts. Likewise, operating a web browser and a search engine to locate appropriate visual resources provided the ideal level of challenge and interest. Perhaps the most difficult task for her to master involved saving and retrieving files, although with practice she did manage this task very well. Documents produced by the student provided the perfect vehicle for stimulating dialogue between Linda and the teacher and between Linda and her peers. During this initial phase no written text featured in any of the activities. This meant that Linda did not exhibit any anxiety and demonstrated much enthusiasm for the task. It also allowed for a great deal of repetition in the use of the desktop publishing tools while also providing variety, since each document produced by the student was different.

Using the desktop publishing software, the Internet browser and the search engine required the development of basic computer skills such as turning the computer on and off, starting programs along with saving and retrieving files. As the student gained competence in using familiar tools in the program, she demonstrated a willingness to randomly experiment with other features of the software. This experimentation and risk-taking was encouraged. The use of computers for students with intellectual disabilities has often been critiqued on the basis that the emphasis is nearly always on dubious drill and practice software and that this often leads to lower educational outcomes. Although special education uses have been characterized by heavy reliance of drill based exercises, this has not been all that different to computer use in general education.

In 1998 Wenglinski examined national U.S. data and found a dominance of drill and practice classroom computer uses and also demonstrated the lower academic achievement associated with this strategy compared to ICT activities that evoked problem solving. More recently, Meelissen and Drent (2007) reported that most cases of school ICT use were still limited to drill and practice, remedial teaching and word processing. They criticised this approach as being not innovative and not very challenging to students. In contrast, Linda's thematic program did not involve drill and practice and afforded the right mix of challenge and opportunities to embed structured but relevant learning. Support for the thematic approach to linking the curriculum with ICT has been provided by Cantu and Farines (2006), whose conclusions emphasized the motivational benefits and the greater incidence of meaningful learning with ICT enhanced thematic approaches.

Access to higher level software was identified as an equity issue by Sutton (1991) who reviewed twenty years of classroom technology research and concluded that disadvantaged students had far less access to more sophisticated software than did more advantaged students, mainly because of the way teachers assigned activities in a differential manner. This leads to advantaged students feeling more comfortable in their post schooling use of computers and their computer dominated work environments, effectively disadvantaging students with intellectual disabilities in terms of their enrolment in computer related higher studies or accessing computer related employment.

Despite the usually mundane uses of computers in schools for drill or typing, enthusiasm for working with computers has been extensively reported in the literature. In relation to students with intellectual disabilities Gardner and Bates (1991) study reported on attitudes to computers in the classroom. A summary of the findings were:

1. Computer work was considered easier than other kinds of school work.
2. Students thought they progressed more with their work when they used computers.
3. Students believed that the computer could effectively teach them things.
4. Students enjoyed using computers in school.

5. Students believed that it was easy to know when they got things right on the computer.
6. Students felt intelligent when they used computers.
7. Students wanted to use computers more during the school day.
8. Students believed that using computers represented work not play.
9. Computer work was considered more enjoyable than other kinds of school work.

More recently Carey, Friedman and Bryen (2005) reported on the high level of interest that students with intellectual disabilities have in ICT and commented on the great promise that new technologies such as computers, cell phones, PDA's and the Internet held for these students. Skillen (2008) conducted a study that investigated the use of Internet based media resources and concluded that the technology allowed motivation to be enhanced and that there was an increased willingness of students to complete tasks and engage in discussions. During this program, Linda appeared to be highly motivated by the access to digital resources and also demonstrated a high level of task completion and an eagerness to discuss visual features of the animals or their environment.

An important goal of any program for students with intellectual disabilities should be to equip them with essential life skills that will enable people to make the transition from school to living and working productively in the community. Indeed, this is mandated in the Individuals with Disabilities Improvement Act (IDEIA, 2004), which requires schools to make provisions for students to access employment, post-secondary education, post-school activities, general participation in the community and successful independent living. Competence with new technologies has become an essential requirement in modern society. Lack of engagement due to incompetence or lack of access to ICT resources often causes crippling and impoverishing effects. Carter, Lane, Pierson and Stang (2008) document the substantial efforts that are being made to ensure that students with disabilities build up a set of important life skills so they are prepared for a life involving self-determination. They are concerned that the emphasis on inclusion and access to the general education curriculum might result in less time spent on developing essential life skills. They rightly raise questions about where and how and when educators can provide important life skill training in combination with the meeting of general curriculum demands. Li-Tsang, Yeung, Chan and Hui-Chan (2005) reported that "as modern society relies more and more on IT in daily activities, the poor computer competency of people with intellectual disabilities may lower their level of participation in leisure, functional and vocational aspects of life" (p. 127). Including a strong ICT component in the program for Linda facilitated interest and motivation and linked to the general curriculum but also assisted in the development of critical life skills. Tsang, Lee, Yeung, Siu and Lam (2007, p. 560) investigated the long term effects of ICT training for people with intellectual disabilities and concluded that the benefits "include the enhancement of their daily functions and literacy, extension of social networks, improvement of independency and quality of life, and facilitation of empowerment."

Reading and writing development

Reading and writing were activities that Linda had decided to avoid, and she had developed effective strategies to ensure that she would rarely need to participate meaningfully in formal and traditional approaches to remedial, small group or classroom offerings. Joseph and Konrad (2008) conducted an extensive review of research studies centred on teaching children with intellectual disabilities to write. They found that students with intellectual disabilities have fewer opportunities to write during their schooling compared to students without disabilities and that, despite the advantages of new technology to better enable students with intellectual disabilities to write, this had not improved in recent times. They pointed out that students with intellectual disabilities have extra challenges in relation to acquiring reading skills such as obtaining skills at a slower rate and not easily acquiring or retaining strategies that are typically taught to scaffold writing development. They may also have difficulty generating new ideas or linking prior ideas or concepts to new work. Other areas of weakness may include planning, organizing and monitoring.

Despite these challenges, the review supported the argument that students with intellectual disabilities could learn to express themselves successfully through various forms of writing, particularly when mediated by the infusion of information communication technologies. A key enabling factor was that students needed to interact with language on a daily basis and that educators needed to devise programs that successfully increased the students' engagement with language. They concluded that "teachers of children and youth with intellectual disabilities are encouraged to find ways to embed such instruction into their daily curriculum" (p. 17) with the help of assistive technology and modified programs. Specific strategies mentioned include modeling, providing guided practice, correcting errors and providing ample opportunities to practice. Modifying Linda's program so that writing could be gradually introduced as a component of learning desktop publishing meant that her enthusiasm for working on the computer could be capitalized on since she had not built up a barrier of resistance to this type of work. As she gradually started to use words and sentences in her computer based work, strategies such as modeling, correcting and practicing were escalated.

Although the emphasis in the early stages of Linda's program was not on text but on simply adding appropriate borders, photos and graphics, it did afford an opportunity to discuss features of the jungle animals and their environment. This discussion was later steered towards finding text based information for guiding reading sessions. It was important to introduce this in a nonthreatening manner to dilute any threat of evoking past negative reactions to dealing with text. Since she looked on the computer based work as entirely different to her past literacy building programs, little resistance or negativity was encountered. In a recent review of the literature related to developing reading skills for children with disabilities, Martin, Martin and Carvalho found that positive attitudes to reading are critically important for the successful development of lifelong reading skills. They found that attitude affects motivation and achievement through increasing or decreasing the amount of time spent reading.

The review provided support for a whole language approach to reading for students with disabilities rather than an approach centred on direct instruction and the teaching of complex rules. Approaches based on direct instruction and the building of reading strategies in a de-contextualized way had failed dismally with Linda during her previous years of schooling. They defined a whole language approach as being a child directed model where "students read and write based on whole texts with little explicit decoding instruction. A whole-language curriculum focuses on developing literacy strategies through existing literature. Students discuss the literature and otherwise interact with the text as their basis. The intention is to link classroom experiences, life experiences and background knowledge to gain comprehension of the literature. The whole language philosophy is based on the assertion that literacy develops naturally, in much the same manner as language is acquired" (p. 117). In Linda's case the literature sourced was entirely web based and always associated with the completion of documents within the desktop publishing instruction.

Peer tutoring

A risk involved with Linda undertaking this program was that is was essentially a 'withdrawal' strategy in which she worked separately from her classroom peers. Although the work was undertaken in the classroom and matched the unit of work conducted by her peers, it did involve different work completed by Linda alone with the assistance of the teacher and teacher aides. This method does risk further alienation from general classroom peers. Previously Linda successfully camouflaged her lack of engagement with reading and writing tasks but harboured a high level of frustration and low academic self-concept. The pretence maintained the calm classroom equilibrium but had not led to genuine acceptance by her peers or contributed to building academic competence. As previously outlined, the use of computer based learning engendered an enthusiasm not evident in other types of programs trialled with Linda.

After she worked with desktop publishing for several months, something unexpected occurred that forged a link with her regular class members and revealed a growing respect for Linda's skills in this area. Other students became fascinated with the desktop publishing software, particularly when Linda started producing products with attributes that could not be easily replicated using word processing software. Suddenly Linda was not working alone at the computer on her tasks but often had other students sitting beside her being shown various aspects of the desktop publishing software. Although this turn of events had the potential to interrupt Linda's workflow, the potential benefits included not only greater acceptance by her peers but much needed practice in communication skills. Stenhoff and Lignugaris/Kraft (2007) completed an extensive review of the literature in relation to peer tutoring and students with disabilities with a view to providing a firm evidence base for the use of this practice. They made an interesting point that the "Individuals with Disabilities (Education) Improvement Act" (IDEIA, 2004) requires that teachers employ strategies for teaching and learning for students with disabilities that are supported by clear evidence from research. Their aim was to provide teachers with this evidence base. They successfully documented clear evidence that peer tutoring increased the amount of 'engaged time' in elementary school settings and that peer tutoring did have the potential to increase academic achievement. They stated that it was very important that the peer tutoring was carefully monitored, and this aspect was evident in the progression of Linda's use of peer tutoring. However, they also make it clear that outcomes are enhanced when effective training in the elements of peer tutoring is provided to the tutor prior to tutoring commencing and is continued through the process.

In Linda's case the peer tutoring training was not provided prior to the tutoring occurring since it was not a planned aspect of the activity. Once the tutoring began, it was encouraged and some ad hoc advice was given when difficulties or problems became evident. The elements of training recommended by Stenhoff and Lignugaris/Kraft included establishing clear expectations with tutors, modelling behaviours beneficial to the teaching/learning process, providing space for tutors to role model and practice tutoring, teach techniques for reinforcement in delivery, to model corrective feedback strategies, to practice and understand performance measurement strategies, to teach feedback techniques, to learn about problem solving scenarios and to match tutors with tutees. Linda's 'just in time' tips on tutoring did not include all of these elements, so despite the success of this technique in the program, it may have been further enhanced by the provision of structured training. Bond and Castagnera (2006) argue that peer tutoring and other forms of peer support are an underutilised resource in the successful inclusion of students with disabilities in general education. They contend that it is a cost-effective approach at a time when adequate educational funding is not always forthcoming and that involving students in various types of peer tutoring assists to forge a classroom community—a community based on respect for diversity. Bond and Castagnera concluded that "optimally, in cooperative classrooms, students with disabilities are not the only ones to receive help, they provide help as well" (p. 228). Through this program, Linda became the person who could offer help with desktop publishing.

Development of oral communication skills

An important goal of the program was to increase the opportunities for verbal interaction between the student and the teacher and teacher aides. Peer tutoring also afforded an unexpected opportunity for increased student to student verbal interactions. Many researchers have emphasised the importance of the development of life skills to aid in the transition between school and community living for students with intellectual disabilities. A key life skill is competency in verbal communication, so this was given due emphasis along with the development of writing and reading skills. Prior to the introduction of the program, little opportunity existed for Linda to practice verbal communication skills. The use of a theme is in accord with the findings of Myers (2007), who recommended an integrated approach involving language, literacy and ICT—all being taught simultaneously. She argued that communication skill development in students with intellectual disabilities would be better supported by a thematic approach since learning in this model occurs through association rather than attempting to teach or learn a set of disconnected facts. The approach of integrating oral communication, literacy and technology skills was endorsed and supported by Myers's study. She also recommended that teachers maintain high expectations for oral communication, and this sentiment was echoed in the research conducted by Balandin and Duchan (2007) that identified this factor as the primary barrier to full inclusion and communication development. Balandin and Duchan edited a special edition of the *Journal of Intellectual & Developmental Disability* that concentrated on the importance of developing effective communication skills in people with intellectual disabilities. Ferguson (2008) implored teachers facilitating inclusion to move "away from the traditional didactic teaching format where teachers lecture, describe and explain and sometimes question while students are expected to listen, quietly and passively, until instructed to say or do something" (p. 114). Students such as Linda also need to understand how to use a variety of communication methods to express themselves. Linda located resources on the Internet to support her theme, discussed these resources with the teacher and teacher aides and displayed her finished product and discussed ICT techniques with her peers. A logical extension of this would be to publish some of the products or thoughts in the form of blogs, podcasts or via the creation of websites.

Development of metacognitive skills

An explicit tactic used in the program from the outset was a deliberate attempt to prompt Linda to verbalise and reflect on successful strategies that she used to solve problems and complete work. This was considered to be an essential exercise to make the importance of cognitive processes clear to the student. Unlike many elementary students who learn to read using narrative texts, Linda's reading concentrated on informational texts. McTavish's (2008) research stressed the importance of developing metacognition, the process of thinking about one's own thinking, in order to facilitate competence in learning to read informational texts. Linda was encouraged to use 'think aloud' techniques in addition to adults working with Linda asking questions and providing prompts that would result in reflecting on thought processes to assist in the development of metacognitive awareness.

McTavish found that using metacognitive training with informational texts worked well if the tasks were authentic, such as being linked to a theme or an overall problem rather than employing a set of comprehension questions. Moreno and Saldana (2005) argued that "metacognition and self-regulation are processes extremely relevant to the education of persons with intellectual disabilities. They play a central role in specific limitations, such as outer-directedness and lack of strategy transfer, and are related to desirable educational objectives such as self-determination" (p. 341). They listed abilities such as choosing strategies, identifying and establishing aims, courses of action to achieve them and reflecting on the success of strategies as all being associated with metacognition. They tested an intervention involving a computer program that aimed at training students with intellectual disabilities to develop metacognitive awareness. Students undertaking the program did demonstrate initial gains and had maintained the gains when re-tested 6 months later. Although Linda's metacognitive training was not automated via computer software, it was built around similar principles and strategies that were employed by the software designers. A key goal of Linda's program was to shift her away from pretending to learn to a different model where she was genuinely becoming a self-regulated learner engaged in authentic tasks. Perry, Hutchinson and Thauberger (2008) identified the key components of self-regulated learning as being metacognition, motivation and strategic action and pointed out that self-regulated learners are more successful in and beyond school. The strategies employed during the program were in recognition of the importance of metacognitive awareness as a key feature of self-regulated learning which in turn leads to self-determination.

Building positive self-perception

During previous attempts to engage Linda in activities to improve her reading, writing and other forms of communication, it was obvious that her stubborn resistance and negative attitude towards using text was the most significant barrier. This resistance was thought to be linked to past frustration and failure associated with learning to read and write, which led to a very low level of academic self-concept with long-term flow-on effects to her overall self-concept. It was recognized that building overall self-concept is a long term goal and that individual education interventions are not usually effective in raising overall self-concept. However, one important component of self-concept, academic self-concept, could be effectively targeted. Li, Faitam and Man (2006) wrote that that academic self-perception was one of three key types of self-concept identified by young adults with intellectual disabilities as being the most important to them. Other educators have discussed the possible link between academic self-concept and academic performance, but Stringer and Heath (2008) demonstrated that there was not a simple causal link between the two. Simply increasing academic self-concept by always giving students positive reports or providing inflated scores does not lead to increased academic performance. In their study no link between gains in academic self-concept and academic performance could be found. However, they did report on the benefits of increasing academic self concept for the overall well-being of the student.

In Linda's case low academic self-concept over time had resulted in a refusal to participate in activities that may have offered her a change of making genuine gains. Improving Linda's sense of overall well-being was a worthy goal of the program, but it was also anticipated that increasing her sense of genuine competency would help alleviate the barrier of negativity that stood between her and academic success. Ensuring students' emotional well-being may be even more important in non-segregated settings as Cooney, Jahoda, Gumley and Knott (2006) reported that students with intellectual disabilities in mainstream settings reported significant additional stigma than their segregated peers. They pointed out that "negative treatment reported by children was a serious source of concern and there is a need for schools to promote the emotional well-being of pupils with intellectual disabilities" (p. 444).

Outcomes of the intervention

A risk of this program was that, as the rigor of the work and the expectations escalated, Linda might lose her enthusiasm for working on the computer. During the 12 months of the program implementation, this did not occur. Perhaps the growing recognition of Linda's expertise in desktop publishing by her general education classroom peers provided a boost to the maintenance of high levels of motivation. Evidence of this recognition by classroom peers included positive verbal feedback from other students and a growing demand for Linda to act as a peer tutor in desktop publishing.

As more and more text was introduced in addition to the earlier emphasis on photos, graphics and borders, Linda did not revert to her previous tactics of avoidance and demonstrated a willingness to seek out and use text based information. She was also willing to express herself by typing text into her theme based documents. Reading and writing text became part of the everyday educational experience for Linda, resulting in gains that were not quantified through formal testing but evident to the classroom teacher and teaching assistants through observation and subsequent notes on improvement. A marked positive change occurred in Linda's attitude to text, although she still exhibited a certain amount of negativity to reading standard books or using handwriting. Often students with intellectual disabilities have problems with fine motor coordination, and this deficiency was noted in her guidance assessment records. This would explain why typing text would be very appealing to students with fine motor problems as the desktop publishing features can be used to enhance the text with different font styles and colors and the finished product looks neat and professional. In some cases this is in complete contrast to the quality of handwritten text when fine motor skills are minimal. Participating in this program appeared to facilitate gains in computer competence, reading and writing competence, attitude to academic work and communication skills.

Would the program work with a larger group of students with intellectual disabilities?

Results from the program undertaken by Linda over a 12 month period in her fifth year of schooling pointed to the advantage of strategies that might be useful for other students with intellectual disabilities. These advantages included:

» Bypassing prior negative perceptions of school work by presenting ICT based work as new and exciting

» Overcoming frustrations associated with poor fine motor development that may hinder some students through the common use of handwriting in schools

» Building up very important life skills for students with intellectual disabilities associated with basic computer and Internet use

» Increasing motivation through the use of self-selected themes

» Providing students with an area of competence (desktop publishing) that would be potentially higher than their general education classroom peers

» Providing a platform for the gradual escalation of engagement with text

Although the program used for Linda evolved over the year in an ad hoc manner, trialing the program with a larger group offered an opportunity to formalize and structure the program. For example, the peer tutoring component would be introduced after formal training in effective ways of peer tutoring. A more rigorous regime of data collection would be used to build up an evidence base for other teachers wanting to replicate the program or use particular components or strategies.

Choosing a group of students

Three schools in the town had students with mild to moderate intellectual disabilities enrolled and included in general education, completing their fifth year of schooling. The three schools included the same school where Linda was enrolled, another school with one student and the third with the remaining three students.

Due to the success of the program with Linda, all three schools were keen to be involved in a trial using similar tactics. Unfortunately the student at the second school transferred to another city, leaving four students across two schools with Robyn, Rebecca, Mark and Mathew as the participants.

Differences in the second program

In the first program the activities were designed to suit a student who demonstrated a high level of interest in computers and a very negative perception of traditional classroom pedagogy. In contrast, it was not known if the new group of students had any interest in computers or if they had negative perceptions of their regular work. All four students' records indicated poor fine motor skills and a history of failure with reading and writing. The new program would require coordination between a larger group of teachers and teacher assistants, along with training sessions for teaching staff in desktop publishing, use of the Internet and file management. It was not assumed that teachers and teacher aides would necessarily have these skills. A procedure and guidelines for training the students in peer tutoring techniques would need to be devised. Appropriate instruments for measuring key outcomes from the program would need to be selected or designed.

Measuring results: An evidence based approach

Pre-tests in reading, self-concept areas (including computer self-concept and attitudes) and computer skills were administered to the four students involved in the study prior to the commencement of the formal intervention. Test results acted as "baseline" measures for comparative purposes at the end of the intervention. In addition, 15 minute interviews using structured, open-ended questions were undertaken with the participants individually to provide information about the students' attitudes to computer work and peer tutoring and to give the participants an opportunity to voice any concerns that they may have had about implementation of any of the intervention steps. This information was used to modify the timing of the intervention steps and to ensure that an appropriate amount of scaffolding was given.

The formal intervention commenced, and during the two-year implementation time frame, data collection took take place on a continual basis. Specific data collection tools and techniques are discussed later in this section. Once the intervention was complete and post-test measures of spelling and self-concept areas (including computer self-concept and attitudes) and computer skills were calculated, another round of structured interviews with participants, regular class teachers and specialist inclusion teachers was undertaken.

A group from the regular class members' schools in the same year level (25) was administered a computer skills checklist to determine the average class levels of computer skills at the end of the year, in order to compare them with the study participants' computer skills. One of the participants attended the researcher's school (MSS), whereas the other three attended a different school (GSS).

Quantitative data collected included pre- and post-measures of reading, spelling, computer proficiency, and in the self-perception area: the self-concept subsets of academic, social and physical self-concepts with particular reference to computer self-concepts and attitudes.

Data analysis was framed within two case studies, one of which concerned the student at MSS and the other deals with the three students at GSS. This analysis combined both qualitative data about the students' peer tutoring / communication skills through the use of student and teacher interviews as well as audio and video recording of peer tutoring sessions and quantitative data including results from the test instruments. Combining qualitative and quantitative methods is a strategy that has gained acceptance because of its potential to draw from a wider base of data and to provide richer and more telling data analysis.

The peer tutoring communication abilities of the participants were assessed by triangulated qualitative and quantitative methods, whereas academic outcomes, attitudes and self-concepts were subject to quantitative data collection via test instruments, and the two tests used to assess computer attitudes complemented each other. Triangulation occurred when these were combined with participant interview data and teachers' comments. Complementarity was used across the study to gain information and build up a richer overall picture of the students' outcomes, and triangulation was used extensively to ensure reliability and validity.

The group of students were in the IQ range of 60–75 from two schools in the same general geographical location of a relatively small, rural town. Pre-test measures were taken in reading, computer / desktop publishing proficiency, attitudes and self-concepts. The next step involved the students learning desktop publishing skills to express self-selected interest themes. Desktop publishing was taught in groups and individually by both the author and the classroom teachers. In most cases the instruction was on an individual basis to cater for the different student interests and abilities. The participants were introduced to and taught the features of the desktop publishing package before their regular class peers. This process was ongoing for the entire time of the study and replicated the technique used with Linda.

The next step involved the students in learning peer tutoring strategies and using these strategies to peer tutor their regular class peers in aspects of desktop publishing that they have acquired. As with the development of desktop publishing skills, this was an ongoing process during the time of the intervention. There wasn't a stage during the program when the students had completely mastered desktop publishing or peer tutoring, as one would expect, so training was ongoing.

This was followed by post-testing in reading, computer / desktop publishing proficiency, attitudes and self-concepts. Post-testing of a larger group of general education students in the computer / desktop publishing area was used as a comparison with the group of students with intellectual disabilities. Data collection techniques such as student interview, teacher interview, observation, research diary, audiotape and videotape sessions were used throughout the treatment time to gather data on the peer tutoring process. Audiotaping was useful to determine changes in communication during peer tutoring sessions, for example, the number of words in sentences, but was not appropriate to assess the students' reliance on nonverbal cues. This accounted for the need to videotape some sessions. Approximately two hours of audiotape and one hour of videotape per student was collected at three stages of the intervention: the beginning, the middle and near the end.

Incidents deemed to be 'critical' were documented as a means of illustrating important individual developments pertaining to the outcomes of the program. Reflective analysis was used to determine what incidents were critical or what was not. The author's judgment and intuition augmented by field notes was used to determine the importance of events to illuminate important information related to outcomes of the intervention. The descriptions of 'critical incidents' were based on several sources:

» Direct observation—events observed by the researcher, principals, teachers or para-professionals involved in the implementation of the intervention.

» Interviews—Informal, open-ended interviews about critical incidents were conducted with the classroom teachers, the support teachers, the principals and the participants.

» Documents—Documents involving correspondence between the classroom teachers and parents were read (these were offered on a voluntary basis by the teachers and parents and were not kept due to their sensitive nature—a diary entry was made concerning the correspondence).

» Archival records—Official records were kept at the school regarding incidents. Records of official meetings between school staff, parents, education department officials and the principal of the local special school were retained by the state education authority and viewed by the researcher.

» Self-concept instruments—pre- and post-intervention

Perhaps the crucial criterion for determining critical incidents was that the incident needed to convey an important message or add to the understanding of areas targeted by the program.

Instruments for collecting evidence

Self-perception

The test instruments for self-perception used included the SDQ1 (self-description questionnaire) by Marsh (1990), which assesses four areas of non-academic self-concept and three areas of academic self-concept as well as a self-esteem component.

The clearly defined areas examined by the SDQ1 include physical ability, physical appearance, peer relations, parent relations in the non-academic areas and reading, mathematics, and general school in the academic arena. It has a sound theoretical basis and is backed up by a strong body of empirical evidence indicating that the instrument is valid and reliable.

Another instrument used was the ASK-KIDS self-concept inventory. The inventory includes sections to address how children feel about their social and physical selves and how they feel about specific academic areas such as reading, drawing, numbers and motor activities. Part One of the inventory deals with the academic areas, and responses are concerned with current performance, natural talent, task difficulty, effort needed and future performance. In Part Two the children are asked about friendship, belonging, self-expression, individuality, body image, and appearance. The responses are marked on a graphical scale of faces with various expressions or a graphical representation of increments. This graphical representation was designed with pre-school children in mind but has obvious application to participants or groups with poor reading and writing skills such as the students in this group.

Computer skills and attitudes

The Upper Primary Classroom Computer Attitude Inventory chosen for the project was developed after an extensive literature review of available instruments revealed that few computer attitude inventories had been designed with elementary school students in mind, particularly upper elementary school. The inventory designers determined that three important areas for elementary school children's computer attitudes were:

a) The importance of the computer to the upper elementary classroom,
b) The student's confidence in using the classroom computer, and
c) The students' liking for the computer. These three areas cover affective, cognitive and behavioural domains.

Computer skills checklist

A checklist of computer skills associated with desktop publishing and using the Internet was constructed so that pre-, middle- and post-measures of basic computer competency could be determined.

Reading tests

Three standardised reading tests were chosen—the Neale Test, the Milton Test and the St Lucia test. These tests covered vocabulary, comprehension and oral reading skills and could be administered at appropriate points during the program.

Oral communication

A checklist of communication attributes was constructed to provide a means of comparison in a like situation. The checklist was to be used at various points of the program in a common situation, i.e., peer tutoring sessions. Since communication skills vary in different contexts, it was determined that the checklist would only be used to assess changes in the peer tutoring communication over time.

Results of the program

Although it was not initially known whether the students would respond in a similar way to Linda, it soon became evident that all four students did enjoy working on the computer and all responded more positively to the work. Although the degree of response varied, all the adults associated with the program reported increased levels of motivation and engagement by the students compared to their previous efforts with the regular classroom or segregated remedial work. All four students showed gains in each of the formal measurement scores or inventories. In the critical area of reading, one of the students scored highly enough on the standardised tests to be assigned a formal reading age by the end of the year and the others showed improvement. When the five students exited elementary school (including the first student, Linda) to make the transition to high school, they had all been assigned a formal reading age, despite the overall state average of less than 2% of students with mild to moderate intellectual disabilities being assigned formal reading ages.

An example of the final reading results for Rebecca was:

Figure 9. Rebecca's reading score pre- and post-test comparisons of the Neale and Milton tests

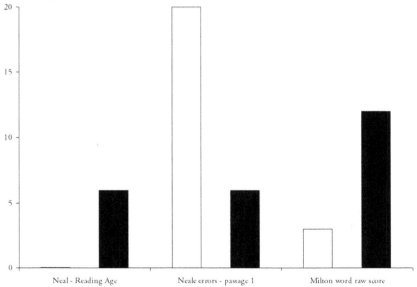

Peer tutoring was a greater success with the larger group due to the early training and knowledge that peer tutoring would be introduced. The peer tutoring steps used in this program were a modification of those employed by Fantuzzo (1992) and colleagues and Marchand-Martella and Martella (1993). The structured steps described by these researchers were altered to suit this particular group of students and to combine the best features of both studies. Their studies demonstrated that peer tutoring is more successful when the participants are shown explicit strategies to implement while peer tutoring and this was the method used during the program. These steps developed for the study were in a brief and easy to understand format and provide evidence of the exact approach used by the teachers or teacher aides when teaching the participants how to tutor other students. As the intervention progressed the participants were reminded about the steps (listed below) and were given assistance if any difficulty occurred. It was noticed during observation and examination of electronic recordings that the students not only needed training in the steps but experience in using the steps with different peers and over a period of time, as they did not successfully master all of the processes until later in the year. A summary of the steps were:

Step 1

After ascertaining that the tutee did not have the skill, tutors described and demonstrated the activity to the tutees, for example, the method used by the desk-top publishing program to change the font type and size.

Step 2

Tutors prompted the tutees to try the activity for themselves. Tutees learnt to use appropriate prompts when needed. For example, "How about you try to put a picture in here now?" is a prompt used by one of the participants during the study.

Step 3

Tutors corrected an incorrect response and provided a further description and demonstration if required. In this instance the teacher/teacher aide modelled this behaviour to the students, followed by the students practicing in pairs with each other before attempting this in a real tutoring situation.

Step 4

Tutors praised tutees for correct responses. This praise was monitored by the teacher or another adult at intervals to ensure that appropriate praise responses were given. Tutors were made aware that inappropriate praise responses should not be made. Praise was appropriate for effort. For example, 'good work' was not appropriate for an incorrect response by the tutee so other prompts such as 'nice try but... ' would be more appropriate.

Step 5

Tutors recorded and collected information on the responses of the tutees in a very easy to use format. This step included filling in a checklist about how the tutor felt the tutee had learnt and responded. This provided opportunities for the participants to practice the art of evaluation, which is a higher order skill often underutilised by children with intellectual disabilities.

Data was collected under specific categories of voice audibility, balance between verbal and nonverbal communication, level of teacher intervention required, length of sentences, appropriate use of praise, success of instructions in teaching the tutee and ability to change direction of the session. The transcripts from approximately 15-minute sessions (2 hrs audiotape, 1 hr videotape) per student were examined as a time series for patterns of change.

An example of the data collection on peer tutoring communication for Rebecca is:

Table 5. Summary of progressive development in Rebecca's communication skills from analysis of taped transcripts

	Early in intervention	Middle	End
Appropriate volume	Volume of speech needs to be increased	Better volume	Volume is fine most of the time—sometimes reverts to softer volume
Communicating with words rather than pointing	No—relies too much on nonverbal prompts	No—still too many examples of nonverbal prompts	Yes—uses appropriate levels of nonverbal prompts
Required teacher intervention (number of times)	10	0	0
Length of sentences used	3.7	4.8	5.3
Appropriate use of praise	0	3	3
Success of instructions in informing tutee	Little success without teacher intervention	Rebecca is using slightly longer sentences and less reliant of nonverbal prompts and is successful in informing the tutee.	Rebecca is continuing to show improvement in framing her sentences to suit the task and the particular tutee.

Ability to change direction of the session if not meeting goals	Very little ability to adjust to the situation of unexpected events or students not following unclear instructions	The research notes record that Rebecca is successful in using different instructions for the same task with different tutees. She has the ability to adjust the instructions to suit the tutee.	Yes

In the area of computer skills the students demonstrated a capacity to learn to use quite complex desktop publishing tools. The learning occurred as part of the composition of thematic documents that became more complex over time and in such a way that was not isolated from the task. All too often, computer skills are taught in isolation and then not retained by students unless they are repetitively used to complete authentic tasks. Importantly, the more complex nature of desktop publishing software did not become a stumbling block for any of the students and they all developed a level of competency necessary to complete their own work but also to provide training to their peers.

An example of the records taken for basic computer skills is:

Table 6. Rebecca's computer skills

Function	Pre-test	Mid	Post
1. Turn on computer	Yes	Yes	Yes
2. Start program	Yes	Yes	Yes
3. Open file	Yes	Yes	Yes
4. Save file	Yes	Yes	Yes
5. Save as (knows difference)	No	Yes	Yes
6. Enter text	Yes	Yes	Yes
7. Change font style	No	Yes	Yes
8. Changes font size	No	Yes	Yes
9. Use bold function	No	Yes	Yes
10. Use italics function	No	Yes	Yes
11. Use underline function	No	Yes	Yes
12. Uses word art or similar	No	Yes	Yes
13. Uses tables	No	No	No
14. Uses shading	No	No	Yes
15. Inserts sounds	No	No	No

16. Uses lines	No	No	Yes
17. Inserts graphics	No	Yes	Yes
18. Integrates functions to make a unified whole	No	No	Yes
19. Is aware of document layout	No	No	Yes
20. Moves graphics	No	No	Yes
21. Uses borders	No	Yes	Yes
22. Uses spell checker	No	No	Yes
23. Shows creativity	No	No	Yes

When reviewing the results of the program for Linda, it was thought that her level of academic self-concept had increased and that the kudos she had received in relation to her expertise and peer tutoring had contributed to the development of overall positive self-esteem. This was based on a gut feeling after watching classroom events and Linda's response and her apparent changing attitude, although no formal measures were taken. The same opinion was maintained during the program with the four students and this was confirmed by the SDQ instrument results for Rebecca:

Figure 10. Self-Description Questionnaire results for Rebecca

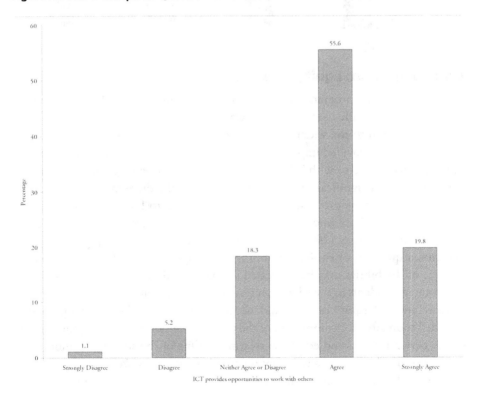

ICT provides opportunities to work with others

Likewise, the ASK-KIDS self-perception (academic) and computer attitudes increased.

Figure 11. ASK-KIDS results for Rebecca

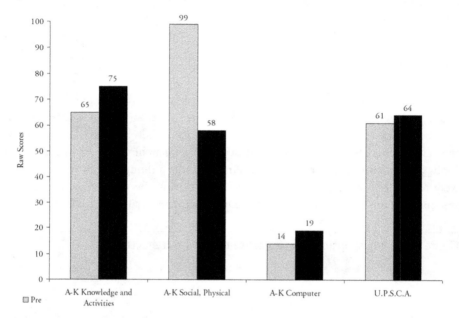

Critical incident: Equity versus equality

Changes to the program and the assessment of outcomes enabled a strong evidence base to be built confirming the academic and social value of the program. Another addition to the wider program involving four students was the recording of incidents that provided important insights into issues surrounding the escalation of inclusion of students with intellectual disabilities into general education settings. Perhaps the most significant incident demonstrated the difference between people who regard equity as providing additional assistance to help students overcome disadvantage and those who see equity as ensuring 'equal' treatment for all students. Dempsey and Conway (2005) wrote that in practice "the philosophy of inclusion becomes a question of equity, or finding ways in which the effects of individual abilities or disabilities may be addressed so that they do not unfairly advantage or disadvantage students in the education system. This is in comparison to the common interpretation of equity as one standard for all" (p. 161). The following incident recorded soon after the completion of the program demonstrated this philosophical divide between equal and equitable treatment. One of the students demonstrated an interest in mathematics, and a component of the program introducing spreadsheets was designed to take advantage this interest. The incident occurred within the context of this activity. A major objective of this program was to develop and implement a framework based on the principles of equity, to involve students with intellectual disabilities in higher order computer tasks. This incident illustrates the difference between philosophies and frameworks based on 'equity' and those concerned with 'equality.' It illustrates the need for 'unequal' treatment for those students at a disadvantage in our education systems.

As part of the intervention process, all the adults involved with Robyn's education were encouraged to use computers as much as possible in appropriate ways throughout the various activities that Robyn was undertaking.

Both the classroom teacher (HS) and the Inclusion support teacher (CA) expressed a desire to assist Robyn to take part in the intervention using computers as a means for Robyn to build up essential skills and confidence, so that she could be seen as an 'expert' in this area by her peers and to ensure that the computer work that she undertook in the program was used and encouraged in the classroom. The computer skills that Robyn built up needed to be made relevant to her regular classroom work and her individual thematic work with the specialist teacher.

The specialist teacher, CA, instigated a program to build up Robyn's life skills of dealing with money. She wanted to include such aspects as counting money, giving change and budgeting and strived to make the activity as life-like as possible along with being relevant and interesting. She also wanted to link it with Robyn's computer mediated language and literacy work by also using the computer for the math work. Robyn had often spoken to CA and HS about her desire to use the computers wherever possible. The activity needed to be a single, meaningful problem that would involve the completion of various small steps and with an overall objective that would be clear to Robyn. Means and Knapp (1991, p. 286) argue that teachers need to focus students on complex, meaningful problems and should avoid breaking things down into unrelated, meaningless skills and that teachers should "embed basic skill instruction in the context of more global tasks."

CA decided that Robyn could gain social skills and life skills in the area of money by making up batches of flavoured cordial and freezing them in plastic cups with a paddle pop stick and then selling the 'icy cups' at morning tea time to the other students. This could lead to the production of advertising posters for the sales and a spreadsheet showing a list of the associated costs and selling prices and profits and some real experience with money rather than standard pencil and paper math activities that Robyn had shown a disinterest in.

CA held talks with Robyn and outlined the idea and found that Robyn was very keen to participate in the activity. Robyn was willing to do the shopping to buy the materials necessary and to spend time budgeting and preparing for the sale. She would also spend time making up the next batch of icy cups and later in the day she would enter the information in the computer and from time to time print out her budget details.

Robyn started the program with enthusiasm and keenly advertised the icy cups to the student community through class visits. During these visits she put up the posters and spoke to the classes about when the icy cups were to be sold, the cost, location and any other relevant information. Creating the advertising posters linked naturally to her desktop publishing skills developed in the main part of her program and provided a form of skill transfer. She also spoke to the whole school assembly about icy cups each Monday morning, reminding other students to bring along some money if they wanted to purchase ice blocks.

As the program went on, the sale of the icy cups was successful in that it offered Robyn the opportunity to speak more frequently to various students in an informal and formal setting as well as giving her authentic practice in dealing with money issues. It also involved continual and relevant practice of computer skills, therefore enhancing the overall strategy of developing her skills in the computer area. It also afforded the opportunity for more complex problem solving skills to be utilised. CA prompted Robyn to consider issues such as making a profit or making a loss, effective means of advertising a product and how to use profits. Lee and Cho (2007, p. 113) define problem solving as "a process of closing a gap between an initial state and a goal state." Young (1997, p. 38) also noted that "evaluating progress towards one's goals" is an important component of higher order thinking. Budgeting towards making a profit through buying and selling was not a familiar concept to Robyn, thus the steps in planning and implementing this venture placed many new cognitive demands on Robyn and monitoring the computer records provided an important means of monitoring the progress of profits and expenditure.

Robyn was often questioned in a way that made her reflect on the thinking processes involved with the success of the icy cup sales or how things could have been completed differently. This technique became evident in discussions between teachers and teacher aides in regard to providing the participants in the program with opportunities to make metacognition explicit. CA reported that she asked Robyn to 'think aloud' the steps taken to achieve various stages of the venture.

As sales increased, problems occurred that needed solving. For example, the lines grew longer and some children became restless waiting to be served. Robyn was encouraged to think about problems and solutions as they arose rather than CA just providing the solution. If Robyn needed extra support and scaffolding to solve a problem, it was given but only at a minimal level to avoid frustration or complete stalling of the process. Robyn solved the problem of the lines by asking two other students to help her sell the icy cups. As the other students were keen to be involved in selling ice blocks during the lunch break, Robyn had little trouble persuading them to help. This led to extra socialisation between Robyn and the two girls chosen.

As the profits built up, Robyn decided that she would like to go over to the café across the road every now and again to buy lunch and a drink with CA. This resulted in a trip across the road to the café to determine prices of food and drinks and an examination of what food sold at the café and whether it was nutritious or otherwise. After a meal and a drink was chosen it was time to examine the computer budget printouts to determine whether or not the idea was affordable. CA felt that as Robyn's lunches were often sub-standard that she would benefit from a café lunch, and it would be a good opportunity to practise social skills and manners associated with eating at a café or restaurant.

CA approached HS about the idea and gained somewhat reluctant approval for the occasional lunches to occur. HS seemed to feel that the activity favoured Robyn at the expense of other students who did not have these privileges and that other students sometimes were involved in fundraising for school activities such as school camps but not to raise money for their own benefit. CA pointed out that Robyn had a somewhat underprivileged home life and her intellectual disabilities meant that educational programs for students with special needs sometimes require providing extra opportunities than would normally be the case. CA felt that as some students start at a disadvantage, extra assistance is sometimes necessary to help students with intellectual disabilities to gain skills and acceptance. HS agreed but did not seem to be fully comfortable with the idea. CA sent a letter home to Robyn's guardians about the idea. They responded by fully agreeing to the program and expressed their satisfaction with Robyn's enthusiasm for the activity.

Robyn continued to sell the icy cups, all the time building up skills in dealing with money, budgets and ways of recording and using information on the computer via spreadsheets. Robyn's helpers continued to work happily with her and engaged in more dialogue. They were also keener to socialise with Robyn at lunchtime and morning tea (this was gleaned from notes taken by CA after discussion with the helpers and observation of their playground behaviour). Every now and again Robyn and CA would enjoy a meal together at the café. As Robyn did not live with her mother, she began to see CA as almost an adopted mother with Robyn confiding in CA and learning much from CA about social skills as well as developing her academic skills.

A close nurturing relationship developed between CA and Robyn. As time went on and the end of the semester drew near, the profits from the icy cup sales exceeded the amount expended on the occasional lunches. Robyn noticed this on the spreadsheet printouts and discussed the problem with CA. Robyn thought that the extra money could be spent on an activity involving the other two girls who had diligently helped sell the icy cups during lunch breaks. Robyn decided that she would like to pay for an afternoon of bowling at the local bowling alley with the two helpers. CA approached HS, who reluctantly agreed after expressing her disquiet about 'extra privileges' and once again CA contacted the parents of all three children who all agreed that an afternoon of bowling would be excellent for the children to undertake.

CA particularly liked the plan, as she had often had to work with Robyn on the development of gross and fine motor skills and saw Robyn's enthusiasm for bowling as being very positive. At the local alley they provided special guards for the children so that the balls could not go in the side alleys. If the ball was going out it would hit the guard and still knock down some pins. This was a good bonus, as standard bowling alleys could prove frustrating to Robyn if the ball continued to go out. The other two girls were overwhelmed with Robyn's decision to offer them an afternoon of bowling and were obviously very pleased with Robyn's generous nature. This further cemented the improving relationship between Robyn and her class peers and the afternoon bowling session proved to be a great success. Robyn continued to work well on other areas of the computer intervention such as desktop publishing, peer tutoring and maintaining e-mail links with friends at her previous school.

The activity continued on a regular basis until partway through the second semester when CA gained a position as Advisory Visiting Teacher in the area of Autistic Spectrum Disorder. This marked the beginning of serious problems that would ultimately seriously disrupt Robyn's education. When CA left the school to take up her new position, a temporary replacement was found to continue work with the children at the school with intellectual disabilities. CA introduced the new teacher (BW) to the children and reviewed the Individual Education Plans. She discussed various children and the way the programs worked with a view to enabling a smooth transition period. CA outlined Robyn's work with computers and money skills in relation to the icy cup program. BW was keen to continue the activity and spoke to Robyn about it. Robyn confirmed her keenness to continue the program, and it was agreed that the program would be maintained.

BW continued the work with Robyn but noticed a growing opposition to the icy cup activity by HS, who actively complained to other staff members about the inappropriateness of "special privileges" given to Robyn and that "other students would not be able to start selling things and then to use the money for their own purposes." Steps were taken to explain the objectives of the intervention to HS and how 'equity' was different to 'equality' and that Robyn may well be in need of 'extra' attention or privileges not given to other students. HS maintained her opposition to 'special treatment' given to particular students. Her statement was that "all students should be treated equally and fairly." HS exaggerated the number of times that Robyn used the money for the provision of lunches when discussing this with other staff and complained that Robyn's guardians no longer sent any lunch with Robyn because they wanted to rely on the icy cup money. This was not supported by the evidence, as BW checked with Robyn and found that lunches were provided as usual.

BW was not as persuasive as CA, and HS seemed to want to take advantage of the changeover, recognising that BW felt insecure in her new position and was not as familiar with the children's needs or the history of various activities. HS appeared to never really be totally in favour of the icy cup program and seemed to use the situation to try to put an end to the activity despite it being included in Robyn's Individual Education Plan (IEP). HS announced that Robyn would no longer be using the icy cup money for her own purposes and that the program was taking away 'too much valuable school time.' The activity was to be stopped and the money that Robyn had accrued would be put towards the school camp that was to take place in the next few weeks.

This episode illustrates the most extreme fallout from HS's philosophy of equal treatment for all students, even if equal treatment meant some students were left behind. Despite this belief, HS was willing to trial the individualised program of desktop publishing with Robyn and did agree that participating in the program had resulted in social and academic gains but nevertheless harboured a great deal of uneasiness about offering students individualised programs that are more resource intensive than dealing with a class as a homogeneous unit.

Conclusion

These two case studies illustrate the potential that ICT offers educators in facilitating genuine inclusive classroom practices for students with intellectual disabilities. Three clear advantages afforded by the use of ICT in individualized programs include the bypassing of prior negative perceptions of traditional classroom work and/or remedial reading programs, providing essential life skills and overcoming problems associated with fine motor coordination. Peer tutoring by students with intellectual disabilities of their general education peers can capitalize on the development of ICT skills and open lines of communication based on mutual respect associated with competence.

Programs that are successful in assisting students with intellectual disabilities to gain academic and social competence in inclusive settings are currently of great value since the policy and legislation of a great number of countries either mandate or strongly recommend a shift from segregated to mainstream settings. Countries that have not yet introduced policies and legislation may need to follow suit to conform to international law which follows recent policy enacted by the United Nations. In the United States, Bruns and Mogharreban (2007, p. 229) pointed out that "due to policy changes, increasing numbers of young children with disabilities are attending inclusive programs. Professionals in these settings need to have corresponding beliefs and skills so that young children with disabilities can experience success with their peers." These policies are apparent in the standards of the National Association for Education of Young Children (NAEYC), the Division for Early Childhood (DEC) and the Council for Exceptional Children (CEC). Supporting U.S. legislation includes the IDEIA, the extension of the Individuals with Disabilities Education Act, and on the international scene the Salamanca Statement from UNESCO in 1994 and the Convention of the Rights of Persons with Disabilities introduced by the United Nations all provide policy and legislative impetus to non-segregated education (Kayess & French, 2008).

Outside the United States similar legislation and policy is common. For example, in Australia the Disability Discrimination Act (1992) and the Disability Standards for Education (2004) mandate greater levels of inclusion (Forlin, Keen & Barrett, 2008). They explain that "throughout the past two decades there has been a move away from educating children with disabilities in segregated schools to adopting a more inclusive approach with mainstream schools in all jurisdictions in Australia" (p. 251). Countries that have largely dismantled segregated schooling and have 80% or more of students with intellectual disabilities include Norway, Greece, Portugal, Spain, Estonia, Lithuania, Luxembourg and Italy (Ferguson, 2008). In Zimbabwe, Israel, South Africa and the Ukraine, policy documents support inclusion but have a lot of progress to make before inclusion will be common (Szecsi & Giambo, 2007; Eloff & Kgwete, 2007).

This shift in policy and legislation is important not only to support the human rights of people with intellectual disabilities but to act on the evidence that academic and social outcomes are superior in the least restrictive environments within general education (Pierson & Howell, 2006). Despite the advantages of inclusive education, process towards full inclusion of students with intellectual disabilities has been slower than expected in the United States (Smith, 2007; Ferguson, 2008) and in Australia (Grace, Llewellyn, Wedgwood, Fenech & McConnell, 2008). Despite the inevitability of increasing numbers of students with disabilities being placed in general education settings, recent reports of debate and polarized opinions on inclusion still persist (O'Rourke & Houghton, 2006). One reason for this continued debate is the recognition that students with intellectual disabilities are more difficult to support in regular classrooms than students with other forms of disabilities (Alghazo & Gaad, 2004; Englebrecht, Oswald, Swart & Eloff, 2003). Studies such as these have demonstrated high levels of teacher stress when educators are unsure of practical strategies to engage students in interesting work that is not met with great resistance. The case studies outlined earlier in the chapter demonstrated how self-selected thematic work, infused with technology, can assist in including students with intellectual disabilities in meaningful activities that can lead to successful improvement in academic and social competency.

Advocates argue that a prerequisite of successful inclusion is a positive attitude towards the inclusion of students with intellectual disabilities and that these attitudes should be built up during pre-service teacher preparation and while teachers are practicing via professional development training. Teachers argue that there needs to be more specialized training provided during their pre-service years. This training needs to focus on the particular needs of students but also needs to ensure that teachers enter the profession well equipped to use ICT throughout the curriculum and to use it in innovative ways rather than employing drill and practice software as a type of babysitter or time filler. It is disappointing to see the recent studies confirming that classroom use of ICT is still dominated by drill and practice software and typing of previously handwritten documents despite numerous studies that demonstrate the harm caused by the overuse of these approaches. Successfully including students with intellectual disabilities requires a shift from a teacher centered to a student centered approach, and this shift from didactic pedagogy would benefit all students.

In addition to familiarity with successful approaches as outlined in the case studies along with adequate training and professional development, teachers have identified reasonable class sizes, adequate support staff and adequate material resources as enabling factors (Eloff & Kgwete, 2007). Forlin, Keen and Barrett (2008) also emphasize the importance of increasing opportunities for collaboration between school staff and gaining the support of the whole school community. Teachers have also raised the issue of extra time needed to facilitate the creation of Individual Education Plans, to develop and modify programs and to fulfill the extra demands of regular reporting mandated by increased quality control and accountability paper-based procedures. Educators need to be supported systemically to be able to realize the potential of ICT infused, innovative and motivating programs that improve outcomes for students with intellectual disabilities and support the ideals of inclusion.

. . .

PART 3

Assistive Technology and Information Communication Technology

WITH JANE BUSCHKENS

New advances in assistive technology such as Schwartz's (2006) work involving monkeys controlling prosthetic limbs via their thoughts capture the imagination and sound as if they belong in a science fiction novel. In 2006, Schwartz, Cui, Weber and Moran (p. 205) reported that "brain controlled interfaces are devices that capture brain transmissions involved in a subject's intentions to act, with the potential to restore communication and movement to those who are immobilized. Current devices record electrical activity from the scalp, on the surface of the brain, and within the cerebral cortex. These signals are being translated to command signals driving prosthetic limbs." In 2008, Velliste, Perel, Spalding, Whitford and Schwartz reported in *Nature* successful experiments, where monkeys' use of the artificial limbs became successful proxies for their natural limbs. The complexity of the movements and the ability of the monkeys to go beyond the experiment parameters to use the limbs in creative ways surprised even the researchers. In one example the researchers recorded that the "monkey moved the arm to lick the gripper fingers while ignoring a presented food target and sometimes used the gripper fingers to give a second push to the food when unloading." These behaviours were not task requirements, but emerged as new capabilities were learnt, demonstrating how the monkey used the robot arm as a surrogate for its own.

Previous experiments had involved very rudimentary movements whereas the monkeys in the recent experiment controlled the arm in different planes and rotations as well as controlling gripper fingers. The researchers see great potential for this technology to assist in the development of human to robotic limbs via similar interfaces. They concluded by claiming that "we have expanded the capabilities of prosthetic devices through the use of observation based training and closed loop cortical control, allowing the use of this four-dimensional anthropomorphic arm in everyday tasks. These concepts can be incorporated into future designs of prostheses for dexterous movement (p. 4)." Assistive Technology (AT) can be as complex and futuristic as this example or it can be the use of a simple technology that makes a great difference to an individual. It can also be an overarching consideration in planning buildings and other environments to suit all types of people with abilities and disabilities. This chapter will define the concept of AT, give an overview of AT devices and their uses and examine current and future trends.

What Is Assistive Technology?

Simply put, assistive technologies *assist* people with disabilities to function better (Scherer, 2005). In extreme cases, life itself may depend on the use of an assistive device such as a pacemaker or respirator (Scherer, 2005). Beyond mere survival, however, assistive technology (AT) can provide the key to optimal performance for disabled persons in almost every sphere of life (Cook & Hussey, 2002). Within their adapted human performance model, Cook and Hussey (2002, p. 44) view assistive devices as "extrinsic enablers" that can compensate for deficits in sensory, motor or processing capabilities that are essential for the accomplishment of an activity. Thus, they see assistive technologies as augmenting, or substituting for, intrinsic abilities. In this deficit model, assistive devices are seen to enhance strengths to counterbalance disability, provide alternative means of accomplishing a task or, possibly, allow a task to be bypassed altogether (Lewis, 1993, cited in Watts, O'Brien & Wojcik, 2004). Thus, the user of AT is rendered able to perform activities that would be difficult or even impossible without such assistance (Bryant & Bryant, 2003).

An assistive technology device has been defined as "any item, piece of equipment, or product system, whether acquired commercially off the shelf, modified, or customized, that is used to increase, maintain, or improve functional capabilities of individuals with disabilities" (Technology Related Assistance for Individuals with Disabilities Act of 1988). This has become the internationally accepted definition, although it has also been criticized as being so broad that almost any item could potentially be considered to fall within the AT category (Edyburn, 2004) so that exactly what constitutes assistive technology still remains contentious (Watts et al., 2004). While the term has typically come to be associated with relatively high-tech equipment and computers (Ashton, 2000), assistive technology devices can and do range from no or low-tech items such as a large grip pencil to high-tech tools such as a voice output communication aid (Merbler, Hadadian & Ulman, 1999).

Rahimi (1981, cited in Scherer, 2005, p. 76) has proclaimed that

"In a world where human beings and the machines they command have the power to control the quality of life, handicapping can only be the result of failure to properly apply technology or the neglect of its development." Indeed, assistive technologies have been described as "world-openers" (Scherer, 2005, p. 32) and the application of technology to everyday tasks, "transformational" (Hitchcock & Stahl, 2003, p. 49). It is predicted that, for those with disabilities, rapid technological advances herald a new era of increased opportunities for productive and independent lives and for much enhanced social integration (Lahm & Sizemore, 2002; Lee & Vega, 2005). Indeed, technological advances combined with an increasing awareness of the potential of AT are considered likely to provide the basis for a more level playing field for those with disabilities (Leung et al., 1999). AT devices have also been labeled the greatest equalizers for students with disabilities and, as such, powerful drivers of effective inclusion (Michaels & McDermott, 2003). However, the recent exponential rate of technological development has led to both great excitement and great confusion surrounding AT use (Ashton, 2000). With estimates of at least 20,000 assistive devices on the market, and about 1000 more developed each year, one of the key issues for AT is finding the most appropriate device to match the individual characteristics and unique needs of each user (Scherer, 2005). In particular, assistive technology pedagogy, as a new and relatively undeveloped discipline, has attracted much debate over best educational AT practice. (Watts et al., 2004)

Overview of Assistive Technology Devices

Given the number of devices available, the numerous possible applications of each and the rate of change in the field, to become fully cognizant of every possible AT option is clearly an unattainable goal. However, an examination of some common types of assistive technologies, both low and high tech, can provide insight into the enormous potential of AT devices. Such an overview can also help to elucidate the barriers to participation and inclusion that first prompted development of assistive devices and provide a basis for exploring some of the current challenges in the AT field. Accordingly, it is useful to apply Scherer and Galvin's (1996) framework for categorizing AT devices by purpose. They propose six relevant areas for consideration:

1) personal and self care; 2) seating and mobility; 3) transportation; 4) blindness and low vision; 5) deafness and hardness-of-hearing; and 6) communication. It is also considered appropriate to include a category related to cognitive processing (Watts et al., 2004). While some of these categories may seem unrelated to an educational setting, AT devices often provide basic functions that may be tacitly assumed pre-requisites for other complex tasks. Thus, understanding of the role of AT needs to begin with a holistic and broad-brush approach (Bryant & Bryant, 2003).

With regard to self-care, AT can provide opportunities to maximise independence and everyday functioning. Items within this category might include simple low-tech objects like Velcro closures on clothing and adapted handles on items such as eating utensils. However, more complex tools such as Environmental Control Units (ECU) would also be included. An ECU allows an individual to manipulate their physical environment, such as turning on a fan, light or other electrical appliance through the use of an alternative switch operation system (Education Queensland, 2004; Scherer, 2005). An input device, such as a keypad or switch, can be connected to the control unit, which then translates the incoming signal and transmits this to a device, perhaps via infrared, radio waves or AC power. The Powerlink is a popular educational example of an ECU which, by allowing activation of equipment such as a tape recorder, can provide independent access to activities as well as enhanced participation (Bryant & Bryant, 2003). Thus, particularly for an individual with severe physical disability, the capacity to activate a simple single switch can provide the potential to control a range of devices from common appliances such as cookware and audiovisual equipment, to many battery operated toys or devices and standard computer equipment (Hasselbring & Williams-Glaser, 2000; Johnstone, 2003). King (1999, p. 109) refers to switches and controls as "pass keys" due to their capacity to overcome barriers at the interface between the AT user and a range of standard as well as specialized equipment. Such a micro control device, if appropriate and if correctly placed, often on a more costly and visible macro AT item, may provide the means to unlocking other AT opportunities (King, 1999). In sum, switches can be an important means of providing a range of choices that are a closer approximation to those available to people without disabilities (King, 1999).

Adapted seating and positioning equipment can also provide better access options by allowing increased physical stability and control, for example, for individuals with cerebral palsy (Bryant & Bryant, 2003). Such equipment can enhance an individual's ability to perform tasks and may be critical prerequisites for educational activities (Blackstone, 1990, cited in Cook & Hussey, 2002). Indeed, correct postural support can be indispensable for using other AT devices such as computers and mobility equipment (King, 1999). Freedom of mobility is proposed to be second only to communication as a human priority (European Commission Information Society Directorate General, 2000) and the vast majority of AT devices are developed for mobility purposes (Scherer, 2005). A wheelchair is a common and often essential mobility device for individuals with physical impairments, perhaps due to muscular dystrophy, for example (Bryant & Bryant, 2003). Depending on the characteristics of the user, a wheelchair may be manually operated either by an assistant or by the users themselves, or it may be motorized and, most commonly, controlled via a joystick (Scherer, 2005). Other items such as mobile standing frames, scooter boards and gait trainers can provide alternative mobility options, and are commonly used with children (Bryant & Bryant, 2003). Mobility devices permit greater independent exploration and social interaction as well as the means to access most environments and gain critical learning experiences (Cook & Hussey, 2002). Technologies that provide access to vehicular travel, such as adapted private vehicles, are also an important AT category for individuals using wheelchairs and can provide both independence and access to wider social experiences (Cook & Hussey, 2002).

Assistive technologies are also employed to compensate for low-vision, where vision loss is such that everyday tasks are impeded but some useful visual ability remains, as well as for blindness, where there is full loss of functional vision (Leventhal, 1996). Products typically address needs related to information access or travel safety. For individuals with some functional vision, options such as screen magnification software can enhance visual ability. For blind people, an alternative, nonvisual mode of information transfer is required, such as the tactile modality provided by Braille. This can be easily accomplished through use of such devices as portable Braille notetakers, which use traditional Braille writer keys but provide the option of auditory reviews or connection to external devices for conversion to text or Braille hardcopy (Hasselbring & Williams-Glaser, 2000). Paperless, refreshable displays may also provide an alternative to a Braille paper embosser (Male, 2003). There are also numerous devices that convey information to the vision impaired user by auditory means, from talking clocks, books and signs to pedestrian signals (Massof, 2003). In addition, Optical Character Recognition (OCR) systems can provide direct access to general print media by using recognition software to convert the text of a scanned document into speech (Hasselbring & Williams-Glaser, 2000; Leventhal, 1996). Since vision is a key sensory mode, enhanced visual contrast or the use of alternative modalities, such as touch or hearing, are vital for ensuring access to content for individuals with vision impairments (Cook & Hussey, 2002).

For individuals who have hearing problems Assistive Listening Devices (ALDs) can help to augment, or substitute for, this core ability (Cook & Hussey, 2002). Such devices can allow access to educational instruction by overcoming key identified barriers in the learning environment such as problems with the interpretation of auditory cues and difficulties maintaining attention in the absence of auditory cues (McFayden, 1996). Examples of ALDs include sound amplification devices, such as hearing aids and phone amplification systems, as well as technologies that minimize background noise such as personal FM systems which create direct links between the user, who wears a receiver, and the speaker, who wears a microphone (Hasselbring

& Williams-Glaser, 2000). It is noteworthy that some devices, such as cochlear implant technologies, which stimulate auditory nerves more directly, are also considered to be ALDs, even though their use may be rather less conspicuous (Moore & Teagle, 2002). The TTY teletypewriter phone system, which sends typewritten words over telephone lines to other TTY machines, is a classic example of a now common AT that was invented to provide alternative telecommunications options for people with hearing impairments (Hasselbring & Williams-Glaser, 2000). Television caption decoders can also provide text in the form of on-screen captions for programs which have been produced with closed captioning (McFayden, 1996). Other useful devices which can compensate for hearing impairment include vibrating or flashing alarm systems which alert the user to important environmental signals, such as the school bell or a phone ringing (Cook & Hussey, 2002). Beyond promoting greater environmental awareness, and thus personal safety, and also improving information access, AT devices for individuals with hearing impairments can play an important role in facilitating social and emotional development through enhancing interpersonal interactions (McFayden, 1996).

Augmentative and alternative communication (AAC) systems and devices provide the means for independence of expression. Given the centrality of language in social settings, this can be critical to psychological well-being (Bryant & Bryant, 2003). Unsurprisingly then, devices for communication purposes constitute a large AT category (Scherer, 2005). Language is such a powerful tool, and AAC devices can furnish users with the critical ability to transmit messages and take part in communicative interactions (Blackstone, 1996). These devices may comprise simple low-tech display boards which a user can employ to indicate messages using words or symbols, but more recently consist of electronic or computer systems which create audible or readable speech output (Blackstone, 1996). The audible speech generated by such AAC devices may be computer generated synthetic output or it may consist of digital recordings of actual speech. Written output may require connection to an external printer or be shown on a display panel (Hasselbring & Williams-Glaser, 2000). The use of communication tools can be critical for many, if not all, interpersonal interactions for individuals with speech-language impairments and/or writing disabilities. In the classroom, AAC tools are likely to be essential for success (Downing, 2000) and can permit students with severe communication disorders to enjoy full participation alongside nondisabled peers (Hasselbring & Williams-Glaser, 2000). While some AAC devices are designed specifically for the purpose of augmentative communication, other communication technologies involve the use of software programs in readily available hardware, such as a laptop computer (Blackstone, 1996). The software user may then employ direct selection of messages from a menu, scanning of alphabet or word displays (visual or auditory) or a variety of symbol systems. The key goal of all AAC tools is to enable active participation such that the user is afforded all the typical benefits of interpersonal communication such as expression of complex needs and wants, information transfer and social connectedness (Blackstone, 1996).

AT devices can also assist individuals with perceptual, cognitive or psychosocial disabilities (Braddock, Rizzolo, Thompson & Bell, 2004; Cook & Hussey, 2002). "Cognitive prostheses" (Watts et al., 2004, p. 43) can provide scaffolding for an individual with limited capacity for cognitive processing tasks such as planning, sequencing and/or remembering (Braddock et al., 2004). With use of AT, students with learning disabilities can be assisted to more effectively and independently manage tasks they might otherwise be unable to complete (Watts et al., 2004). Newer applications of items such as Palm Pilot organizers show promise for helping students with learning disabilities or Attention Deficit Disorder to complete tasks by providing such assistance as step-by-step prompts and reminders about deadlines (Epstein, Willis, Conners & Johnson, 2001; Male, 2003; Riffel et al., 2005). Students with learning disabilities can also profit from products which provide scaffolding during reading and a more interactive style of engagement with the material (Male, 2003). One such product is the Quicktionary Reading Pen which uses optical character recognition combined with speech synthesis. The pen provides a portable and inexpensive support which is used to scan a section of printed text and read it aloud for the user (Higgins & Raskind, 2005).

It should be noted that the use of AT devices as cognitive prostheses is the subject of considerable debate and does not enjoy universal acceptance or understanding amongst educators (Edyburn, 2004). However, it is also noteworthy that physical disability is argued to be a lesser barrier to employment in the 21st century than in previous eras due to socioeconomic shifts toward information management. This shift is simultaneously, however, considered particularly disadvantageous for individuals with cognitive, behavioural or social deficits (Scherer, 2005). Thus, while the question of the pedagogical use of cognitive tools remains to be fully explored, there is increasing interest in the field of cognitive prosthetics. Speech recognition systems have been trialed with students with learning disabilities and found to improve subsequent word recognition, reading comprehension and spelling accuracy (Higgins & Raskind, 2000). Furthermore, students with autism spectrum disorders or emotional disturbances can derive benefit from using augmentative communication devices in order to provide comfort and structure in communicative interactions (Male, 2003). Thus, the applicability of AT devices spans a broad range from low to high incidence disabilities.

Computer-Related AT

There have been dramatic increases in the use of computers and computerized technology among the general population in recent decades (European Commission Information Society Directorate General, 2000). Indeed, computers are credited with increasing overall efficiency and productivity and providing substantial life improvements across all population groups (Lahm & Sizemore, 2002). In the arena of education, use of electronic and information technologies has become ubiquitous (Wehmeyer, Smith, Palmer & Davies, 2004). According to UNESCO, mastery of information and communication technology is now considered a critical literacy, alongside reading, writing and numeracy and, as such, a key goal of education (Anderson & van Weert, 2002).

As has been noted in the previous section, the personal computer can itself function as an assistive technology, providing an alternative means of performing activities such as reading, document manipulation, memory tasks and communication (Vanderheiden, 1996). Indeed, much software can be seen as AT or ICT, depending on your viewpoint (Edyburn, 2004). However, it is also true that many assistive technologies have become necessary due to the advent of computers (Bryant & Bryant, 2003). If, for example, an individual has difficulty seeing a keyboard or screen or their motor skills preclude them from being able to use a standard mouse or keyboard, they will be effectively excluded from using one of the most empowering tools currently available (King, 1999). It is useful to examine computer related assistive devices according to the standard computer functions they can enhance or replace. Modifications or alternatives to standard input devices, such as the keyboard and mouse, as well as to standard output devices, such as the monitor and printer, are often required by individuals with disabilities to allow appropriate access and independence. These may involve hardware or software solutions or a combination of the two (Bryant & Bryant, 2003; Education Queensland, 2004).

Keyboard access can be improved through the use of software that can facilitate its use by providing additional visual or auditory cues or altering the responsiveness of the keys relative to user need (Bryant & Bryant, 2003). Keyboards may need to be more sensitive to touch if the user has less strength or range of movement or be required to be less responsive if an individual has difficulty with motor control (Hasselbring & Williams-Glaser, 2000). Some of these features are available in existing mainstream products, such as the Windows control panel which provides filter keys to allow for slow or repeated key strokes and sticky key features that allow key combinations to be handled stepwise (Hines & Hall, 2000). Alternatively, keyboards with an unconventional layout may be required in order to allow such options as one-handed operation (Vanderheiden, 1996). Some users may need to use a mouthstick, hand-pointer or head-pointer in order to access the keyboard, and for these users the relative positioning of equipment items is likely to be a critical consideration that may require AT mounting solutions (Bryant & Bryant, 2003).

Alternative keyboards may also be programmable and able to be customized for a particular user or task. This may involve reducing the number and increasing the size of available key options to provide better access for individuals with physical, sensory or cognitive disabilities. Products such as the programmable Intelikeys keyboard provide such an adaptable, simplified interface (Merbler et al., 1999). Programmable keyboards also allow keys to perform multiple functions (McCulloch & MacMahon, 1998). Chording keyboards can allow specific keys to represent whole frequently used words and so reduce the amount of effort required by the user (Merbler et al., 1999). Other keyboard alternatives include Braille keypads, illuminated keyboards, alphabetical keyboards (to match communication boards), on-screen keypads and voice activated keyboards (McCulloch & MacMahon, 1998; Merbler et al., 1999).

Alternatives to input via the standard mouse may involve items such as joysticks, which still require reasonable strength as well as multidirectional skill, and trackballs, which consist of a stationary platform with a moveable ball and look rather like an upside-down regular mouse (Bryant & Bryant, 2003). Touchscreens can also provide an alternative input option, with touch sensitive grids aligned with on-screen graphics which allow an intuitive mode of access which is particularly popular with young children (Merbler et al., 1999). Voice recognition is yet another alternative means of input. This strategy involves using a software program to convert spoken words into text on the computer screen. This may be a stand-alone product for existing hardware or developed specifically for use within a particular device. The software will usually need to be trained to recognize a particular speaker and may also require a specific discrete style of input where the user must ensure regular pauses or it may allow continuous dictation (Bryant & Bryant, 2003; Education Queensland, 2004).

Simple, single switches, where the user is required to depress one large pad only, using any body part, can also substitute for mouse input. This can allow access by individuals with physical impairments to many simple software programs employing either a basic cause and effect strategy or the scanning and selection of a matrix of on-screen choices (Johnstone, 2003). Single switches can also be used to activate a switch-adapted mouse, where a modified standard mouse can be used in the standard manner to position the cursor but then the mouse click can be achieved via input from the switch to activate a choice (McCulloch & MacMahon, 1998). Augmentative communication devices can also be linked to a computer to provide an alternative input pathway (Cook & Hussey, 2002). Infrared sensors offer another input option for computer use. A person with severe physical disability might use an infrared pointer to position the cursor and then employ another device, such as a pneumatic puff-and-sip mouth-activated switch, to select the on-screen option (Hasselbring &

Williams-Glaser, 2000). Other available hands-free input devices operate by reading a reflective material placed on the user (e.g., dots on the forehead or glasses bridge). This computer navigation style allows screen icons to be selected by simply gazing at them (Ashton, 2003).

Many conventional computers provide built-in output accessibility options. Built-in accessability features, such as those available in the control panel of Windows, can allow adjustments to be made to the display for screen enlargement, contrast management and the possibility of conversion of auditory outputs into visual signals for users with auditory impairment (Hines & Hall, 2000). Simple adaptations such as a more visible on-screen cursor or enhanced contrast may greatly improve accessability for some users (Vanderheiden, 1996). However, additional modified output options may be required as alternatives to the standard monitor, speaker or printer. For those with low vision or developmental delay, specialized screen magnification may be needed. This may involve selection of hardware that provides a larger, clearer screen or use of software options that allow greater visual clarity or cueing (Bryant & Bryant, 2003). On the other hand, on-screen information may instead need to be transformed into speech through the use of screen-reading software. The option of speech synthesis is particularly useful for blind individuals who require information via an alternative sensory modality (Cook & Hussey, 2002). Specialized printers which produce embossed Braille output also often feature a speech output function (Bryant & Bryant, 2003). Text-to-speech devices may also be valuable for individuals with written language disabilities who are likely to benefit from the addition of auditory feedback (Bryant & Bryant, 2003).

Technology may also be needed to provide support for individuals with cognitive impairment or learning disabilities (Education Queensland, 2004). Personal digital assistants are becoming recognized for their potential to support persons with cognitive disabilities in educational, work-related and daily living tasks (Braddock et al., 2004). These may provide communication and diary functions and can be used for both indoor and outdoor monitoring using wireless connections with scheduled or reactive prompting given as required (Braddock et al., 2004). Specialized computer software programs such as Clicker4 and Intellitalk can provide alternative word processing options which offer auditory support, graphics support, multiple access options and customization for the individual user (Education Queensland, 2004). The use of word prediction software may also be useful for students with written language disabilities, communication deficits or mild motor impairments. When used alongside a word processing program, such software can reduce frustration and increase feelings of accomplishment by allowing students to express themselves without being limited by language or motor deficits (Hasselbring & Williams-Glaser, 2000). Voice recognition software, such as Dragon Naturally Speaking, has also been found to have the potential to reframe the writing experience for students with writing disabilities by allowing a freer flow of expressive ideas (Education Queensland,

2004). The use of hyperlinks, which can offer quick and easy information links, and multimedia approaches, which appeal to multiple sensory modalities, are also seen as having great potential for enhancing the learning experience for many students with disabilities (Hasselbring & Williams-Glaser, 2000).

Accessing the World Wide Web

The Internet has dramatically increased the potential advantages of using a personal computer for people with or without disability. All aspects of life are potentially impacted by Internet use, from shopping opportunities and travel choices to education and information management options (Male, 2003). Individuals with disabilities may benefit more than most from such possibilities (Leung et al., 1999). However, newer communication modes, such as e-mail and online chat rooms, can offer particular benefits for individuals with disabilities. Online communication can provide such affordances as additional time for responding, the potential for compatibility with existing AAC devices and the ability to create dedicated online communities without many of the obstacles associated with traditional groups. In addition, the possibility of increased independence of communication and the potential for a disability-free identity can produce fundamental differences in the nature of relationships established online (Cook & Hussey, 2002).

Thus, the ability of computers to function as tools which support greater independence, productivity and leisure options increases dramatically when used in conjunction with the Internet (Davies, Stock & Wehmeyer, 2001). For those with disabilities, the World Wide Web is considered to have greater potential than other information media because, in combination with AT, it is "potentially tolerant of impairment" (Disability Rights Commission, 2004, p. 1). For those with disabilities, the opportunity to access and send information worldwide may be particularly liberating (Hasselbring & Williams-Glaser, 2000). In fact, even local services, which may be difficult to access physically, may be accessible via a website (HREOC, 2002). The Internet is, ever increasingly, a key information source in classrooms, where it provides the means to expand student learning environments well beyond the school limits (Hines & Hall, 2000). Innovations such as electronic mentoring, where expert volunteers mentor students and become a worldwide helping resource on their subject, and electronic field trips, where unknown destinations can be explored online, are examples of online opportunities for empowerment that can be provided without the need for physical access and may be a particular boon to individuals with disabilities (Male, 2003). Furthermore, other options such as list-servs and bulletin boards, beyond their usual purposes, may be especially useful modes for provision of information on disability related topics or products and as sources of advocacy (Cook & Hussey, 2002).

However, in order to make full use of the World Wide Web to access the range of available information, services and entertainment, people with disabilities must typically overcome numerous obstacles (Disability Rights Commission, 2004). The groups most affected by Internet access problems are considered to be those with blindness, partial sight, dyslexia, profound deafness, sign-language users, those with hearing impairment, and those with physical impairments relating to dexterity (Disability Rights Commission, 2004). Accessibility recommendations and guidelines for website commissioners and developers have been set down by the World Wide Web Consortium (Disability Rights Commission, 2004).

It is recognized that individuals with disabilities require alternative access options in order to unlock content and navigate their way around websites. For example, individuals with hearing impairment or deafness are likely to need online captioning or transcripts of any audio content (HREOC, 2002). For persons with visual disabilities screen magnification and screen readers are likely to be necessary. When linked to speech synthesisers or dynamic Braille displays these can provide alternative formatting of content, such as larger text or auditory or tactile output, as well as alternative navigation strategies (Hines & Hall, 2000). It should be noted that PDF formats present a particular barrier to using such solutions, however, because they do not provide AT legible text. For this reason plain text alternatives are required alongside PDF files (HREOC, 2002). Text equivalents for nontext components, such as graphics, are also indispensable for people with vision impairments (Disability Rights Commission, 2004). Some popular multimedia authoring packages do not yet allow access via screen-readers (HREOC, 2002).

For those with vision, learning or cognitive disabilities, the ease of navigation of a site will be especially critical, as will the clarity of the information presented (Hines & Hall, 2000). Accessible websites are clearly organized with good contrast, simple backgrounds, clear and consistent page layouts and large buttons (Hines & Hall, 2000). They allow both keyboard and mouse navigation and are compatible with other AT devices (Disability Rights Commission, 2004). Yet, most websites are relatively inaccessible and fail to comply with even the most basic standards (Disability Rights Commission, 2004). In a recent investigation, 80% of the websites sampled failed to include appropriate accessibility considerations in their design and had homepages that were not accessible across disability groups, particularly for people with vision impairments (Disability Rights Commission, 2004). These major barriers to the use of the World Wide Web clearly have implications for teaching and learning.

Specialized web browsers which allow more simplified and independent exploration have been proposed as a possible means of increasing online educational and recreational options for individuals with cognitive disabilities (Davies et al., 2001).

For the inclusive classroom, the use of a class webpage is another suggestion for reducing information overload and the potentially overwhelming confusion due to multiple possible navigation options. A class webpage could also help guarantee visits to key sites as well as ensure appropriate access options for all class members (Hines & Hall, 2000). Nevertheless, widespread and rapid developments in the high tech realm of computer and Internet options appear to be the source of both great promise and great challenges for individuals with disabilities (Hitchcock & Stahl, 2003). In the educational context, these challenges appear likely to be critical.

Emerging and Future Trends

Future advances in technology appear likely to herald ever greater possibilities for all members of society, including individuals with a very broad range of disabilities (Ashton, 2000). As technologies become smaller, lighter, smarter and more afford-able, hopes have been raised for a better and easier life for all, and particularly those with disabilities (Braddock et al., 2004; Scherer, 2005). Indeed, the European Commission has identified a multitude of mainstream technological developments and breakthroughs that appear to have great potential for the assistive technology field (European Commission Information Society Directorate General, 2000).

Virtual reality (VR) is an example of such a technological development which is predicted to have multiple possible future applications. VR provides multi sensory and interactive experiences via alternative synthetic 3-D environments (Braddock et al., 2004). While it may involve full-immersion strategies using helmets, bodysuits, gloves and/or headphones (Smythe, Furner & Mercinelli, 1995), it may also be provided in a non-immersive way, such as via simple desktop applications (Braddock et al., 2004). Though developed for a multitude of purposes from game playing to pilot training, VR technology has wide applicability to the AT field (Smythe et al., 1995). It can permit perception that may not otherwise be possible by transferring information across sensory modalities. It can also provide a simplified artificial world for skill development and can be adapted to suit the individual user's needs (Smythe et al., 1995). VR can provide modeling and shaping, via built-in 'coaching' software; realism, through access to otherwise logistically difficult situations; new sensory experiences, such as the capacity to play a sport which may be impossible otherwise; and flexibility, so that learning music may come to involve colour displays as well as tone (Darrow & Powers,1996, cited in Male, 2003).

Virtual reality has also been found to be useful in special needs training for mobility, wheelchair use and social skills (Male, 2003), and it can help to overcome some of the cost, dangers and practical obstacles of real-world instruction (Braddock et al., 2004). For people with intellectual disabilities, VR can assist to develop protective behaviours in a controlled and safe environment. Furthermore, VR can allow a sense of both physical and intellectual risk-taking (Male, 2003) and is seen as potentially highly motivating (Braddock et al., 2004). Thus, it is considered to have great potential for promoting educational participation (Braddock et al., 2004) and is predicted to be a major driver of educational innovation in coming decades (European Commission Information Society Directorate General, 2000). However, the VR field is considered to be in its infancy with research and development remaining costly and time-consuming. Nevertheless it is an arena of great interest and potential. For instance, recent advances using mixed reality, where virtual reality is overlaid on real objects, show great promise for enhancing the transfer of skills from VR training into real-world situations (Pridmore, Hilton, Green, Eastgate & Cobb, 2004).

Robotics is a related emergent technology with great potential as AT (Miller, 1998). Intelligent robotics has a history of solving mobility problems and has been applied to power wheelchairs to enhance their capability. Smart wheelchairs have been designed that not only respond to user controls but also to navigation sensors on the chair itself (Miller, 1998). In this way, user effort can be greatly reduced while user safety is greatly improved. Other newer semi-autonomous wheelchairs can combine voice command or map selection technology with laser-guided navigation of surroundings to provide a safe path to a desired destination (Ronnback, Peikkari, Hyyppa, Berglund & Koskinen, 2006). Other uses for robotics include fetching desired objects from difficult locations and manipulating objects on behalf of individuals with physical disabilities. Robots can also be used as digital "guide dogs" for visually impaired persons and can translate external stimuli, such as light signals, into appropriate sensory channels, such as sound, according to individual need (Miller, 1998, p. 126). Personal robots may also provide future automated personal assistants for individuals with physical disabilities, which may lead to greater independence in all aspects of life (Braddock et al., 2004).

In addition, robotics has been found to have potential for assisting children with autistic spectrum disorders. Robots in the form of animated interactive toys, may act as "socially intelligent agents" and provide a controlled social learning environment (Dautenhahn, 2000, p. 154). It has been suggested that such robotic toys offer relatively safe interactions, encourage proactive play, allow learning to generalize across settings and promote development of more holistic perception and, thus, concept formation (Dautenhahn, 2000). Humanoid robots also have been proposed as a tool for assisting the social skill development of individuals with autism (Robins, Dautenhahn, te Boekhorst & Billard, 2004). Gestural interface technology and interactive robotics, or GIR–T, is another emergent technology currently advocated as a possible way to enhance motor skills and to promote speech and language development (Lathan & Malley, 2001). In the GIR–T system, a robotic toy can be operated by a remote therapist via interactive Internet-based software which allows for programming and data collection. The toy can then be used to facilitate social interaction and allow interactive storytelling as well as facilitate multipurpose activities which simultaneously develop educational, play and therapy goals. These can be achieved in controllable steps and any progress is digitally recorded (Lathan & Malley, 2001). This strategy would appear to have great potential for the remote delivery of specialized programs.

A number of other developing technologies are envisaged to become key future enablers. One promising device is the data-glove, which can translate the full spectrum of human hand movements into digitized data suitable for information systems (Kuroda, Tabata, Goto, Ikuta & Murakami, 2004). For example, individuals who use sign language can wear a glove that will translate their gestures into digital text on a computer screen (Braddock et al., 2004). Such wearable data-gloves also show great promise for future accessibility options to a range of standard ICT devices, such as telephones (Kuroda et al., 2004). Smart rooms are also predicted to be a future option for individuals with disabilities (Braddock et al., 2004). These environments have the potential to supersede basic environmental control units. Smart rooms will allow an integrated system to track environmental and individual data using sensors; monitor specific variables of interest; provide automated functions, such as timed door-locks; and allow programmable or remote prompting. These features are predicted to provide for greater independence particularly for individuals with cognitive disabilities (Braddock et al., 2004; European Commission Information Society Directorate General, 2000). Yet another developing AT device is the "emotional hearing aid" (el Kaliouby & Robinson, 2005). This tool has been designed for children with Asperger's syndrome and may become a future portable assistant for assisting social interactions. Though still embryonic, the device is intended to help recognize facial expressions consistent with particular moods and then provide the user with a suggestion for an appropriate emotional and/or behavioural reaction (el Kaliouby & Robinson, 2005).

Haptic devices represent a new aspect of technology predicted to radically alter current communication and computer interfaces (Caffrey & McCrindle, 2004). These devices provide tactile stimulation, thus utilizing a previously often neglected communication channel. However, this can be challenging due to the difficulty of trying to simulate the vast array of human sensitivities to texture, pressure and other tactile sensation (Marks, 2006). Originally designed for motion sensing in VR and videogames, for those with visual impairments some haptic devices may be transformative (Marks, 2006). Multimodal websites are in the process of development. These sites are designed to include both haptics and speech recognition and are hailed as a major leap forward in Internet navigation options particularly for individuals with visual impairments (Caffrey & McCrindle, 2004). Another haptic product in development consists of a headband which provides a tactile representation of information it receives from an optical viewer on purpose-built user glasses. This provides a tactile map which blind persons can use for navigating (Marks, 2006). Prototypes for sonic way-finding systems have also been developed for devices that provide sonic echo location. These sonar tools can also allow independent navigation by blind individuals by providing an auditory map of the surroundings (Massof, 2003).

Numerous future developments are predicted that will impact on the lives of those with disabilities. Future AT is considered likely to include options such as devices that can integrate communication systems, mobility devices and environmental control units. It is anticipated that direct speech recognition will become commonplace and provide a means to avoid problems at the human-machine interface by obviating the need for devices such as a mouse or keyboard. In addition, computers of the future are envisaged that could be made aware of both situational factors and individual characteristics so they would be able to intuit user intentions from speech or actions and provide support automatically. Yet another potential development is a personal data card or chip which could reduce problems at the human-machine interface by carrying information about personal interface requirements. In this way, any device would be able to adjust its protocols according to individual need (European Commission Information Society Directorate General, 2000).

Increasing miniaturization of many technologies is thought to have huge potential for AT. The more portable or even wearable memory aides, location detectors and imaging or translation devices become, the more user-friendly and manageable they are likely to be. The miniaturization of batteries is also likely to be a future trend that will help improve the portability and usefulness of many devices. The technical convergence of many current media, including the Internet, mobile phones, cable and radio telecommunications with hardware systems is also predicted to lead to more mobile and affordable combined ICT–AT options (European Commission Information Society Directorate General, 2000). While some of these projects may never come to fruition, the AT and IT changes of the last decades would indicate that many will. Perhaps "yesterday's high tech is tomorrow's low tech" (Cook & Hussey, 2002, p. 9).

Access, Use and Abandonment

In an era of technological sophistication and relative affordability, there is an expectation that ICT opportunities have become vastly expanded (Lahm & Sizemore, 2002). However, there is also acknowldgement of a digital divide where individuals from disadvantaged backgrounds are less likely to enjoy equitable access. The available evidence indicates that access to computers and the Internet, both in classrooms and homes, is significantly lower among individuals with disabilities (Braddock et al., 2004). Thus, in addition to low family income, culturally diverse background, and female gender, disability is predictive of limited ICT access and usage (Kalyanpur & Kirmani, 2005). So, while educational participation is recognized as an essential ingredient in successful student outcomes, for students with disabilities, access to educational ICT experiences is often limited (Craddock, 2006). This appears to have important implications given recent special education imperatives to address general curriculum standards in Individual Education Plan (IEP) planning (Puckett, 2004). With increasing use of ICT within education it has been suggested that the digital divide between educational 'haves' and 'have-nots' will widen further unless there is a concerted effort to ensure appropriate AT and ICT access (Leung et al., 1999). The more integral ICT becomes to our society the more critical issues of inequity of access will become (Scherer, 2005). For this reason students with special needs are considered to be at exceptionally high risk of disadvantage with regard to information, communication and assistive technology access compared to their non-disabled peers (Jackson, 2003; Kalyanpur & Kirmani, 2005).

Worldwide, much disability related legislation and policy is based on the United Nations' Standard Rules on the Equalisation of Opportunities for Persons with Disabilities, or StRE (Leung et al., 1999). Given that AT use is seen as having extraordinary potential for improving opportunities for those with disabilities, the necessity for AT consideration is, perhaps unsurprisingly, also embedded in such legislation. One of the most influential examples is the American Individuals with Disabilities Education Act, or IDEA (Ashton, 2000; Puckett, 2004). In Australia, AT use is a key component of the Commonwealth Disability Strategy and is supported by the Australian Disability Discrimination Act 1992. These in turn influence State policies which promote the use of AT for increasing access options for people with disabilities (Leung et al., 1999). Nevertheless, it is widely acknowledged that lack of availability of AT devices is a major impediment for many individuals with disabilities (Scherer, 2005). In the United States consideration of assistive technology in every Individual Education Plan (IEP) for students with disabilities is mandatory (Edyburn, 2004; Puckett, 2004), yet unmet needs remain commonplace (Scherer, 2005). The percentage of American students identified as likely to benefit from AT devices but not able to access them has been found to exceed those that were able to access appropriate technology (Wehmeyer, 1999).

One of the greatest barriers to AT use is cost (Elliot, Foster & Stinson, 2003). This is due to a combination of item expense and lack of available funding (Scherer, 2005). Education is considered to lag behind other areas in the uptake of technology, and in special education, with its relatively small market for product development, the situation is even worse (Braddock et al., 2004). The high cost of many specialized devices, such as AAC devices that may cost thousands of dollars, are often prohibitive (Wehmeyer, 1999). Indeed, many individuals with disabilities must rely on charities to obtain AT items (Scherer, 2005). Lack of information regarding availability and suitability, as well as practical difficulties locating and purchasing equipment can also present barriers to AT implementation (Wehmeyer, 1999). Thus, despite the invention of an enormous number of software and hardware solutions, AT and IT access and usage is limited for many students with disabilities (Kalyanpur & Kirmani, 2005). However, beyond physical access other potential barriers remain. Indeed, purchase of a tool cannot guarantee its use (Merbler et al., 1999). As has been noted previously, any device must be appropriate for the unique needs of the user. Moreover, additional design features of the device itself, such as weight, portability, ease of setup, compatibility with other devices and reliability, will also determine whether device use is considered worth the effort (Scherer, 2005). The level of user need and the existence of alternative options will also play a role, with individuals with higher need and very few alternatives generally exhibiting greater AT usage rates (Scherer, 2005).

Furthermore, Scherer (2005) warns that a belief that technology can be a stand-alone solution does not consider critical psychological factors in the person-machine interaction. She argues that the assumption that device availability of itself will ensure positive outcomes for individuals with disabilities is misguided. Indeed, despite the capabilities of many devices and the promise they may appear to hold, there are high rates of AT abandonment (Hocking, 1999; Scherer, 2005; Wessels, Dijcks, Soede, Gelderblom & De Witte, 2003). In fact, an average of one third of all supplied devices will be abandoned, most in the first year (Scherer, 2005). To explain why this may be so, Scherer (2005) uses the example of a computerized communication device and describes how possession of three personal user characteristics is required for its successful operation—a fondness for computers, the cognitive ability to operate the device and a tolerance for unnatural sounding speech. Scherer (2005) takes issue with the assumption that all people with disabilities will desire to operate computer operated devices and points out that variables such as an individual's knowledge about, interest in and comfort with technology are just as important as the capabilities of the device itself (Scherer, 2005). Thus, familiarity with technology in general can be an important component of access beyond mere availability of specific devices (Peterson-Karlan & Parette, 2005).

Indeed, it is important to consider the likely attitude of the current generation of students with disabilities toward high tech devices. Millennial students are generally seen as likely to be quite comfortable with technology, to feel highly connected to the world, via fast communication channels such as mobile phones and the Internet, and to see technology as a tool for learning. However, there is some evidence to suggest there is much less technology use among students with disabilities (Peterson-Karlan & Parette, 2005), though surprisingly little is known about levels of 'techno-philia' among students with disabilities. This knowledge gap has been described as particularly glaring, given the acknowledged critical value of AT for these students (Peterson-Karlan & Parette, 2005). Consideration of such psychological variables clearly illustrates the importance of what King (1999) has labeled the essential human factors in AT success.

One of the key human factors identified by King (1999) relates to a transparency-translucency-opacity continuum which, essentially, refers to the user-friendliness of a device. Typically, the more complex the device, the more 'opaque' its use is likely to be. Some high-end technologies, such as computerized airplane controls, are clearly designed in such a way that their mechanism of operation is neither obvious nor instinctive. Correspondingly, many assistive technologies may be overwhelming in their complexity or their operation may be counter-intuitive (King, 1999). The perspective of the user is critical for good design and optimal outcome (Cook & Hussey, 2002; Wessels et al., 2003). While their non-disabled peers might exhibit a high affinity for technology, so that "it's transparent to them" like the air they breathe, little is known about the characteristics and use patterns of the same generation of students with disabilities (Peterson-Karlan & Parette, 2005, p. 28). Attitudes toward technology, ease of use and difficulty of learning will doubtless play a role in willingness to use an AT device (Merbler et al., 1999). Nevertheless, those in receipt of even the most well-designed and appropriate tool will still require sufficient demonstration, training and practice for successful adoption and use (Leung et al., 1999; Wessels et al., 2003). It is considered essential that the recipient is not left to intuit the use of an AT device (King, 1999). Moreover, it is important to remember that opacity is truly a matter of perspective, and a potential AT user's abilities, interests and previous experiences and training must, undoubtedly, be key considerations (King, 1999). As King (1999) contends, an opaque device is intimidating and likely to be left in the cupboard.

Learned helplessness has been identified as another important human factor in the success or failure of AT implementation (King, 1999). The experience of repeated failure to be able to use a technological aid, in addition to daily limitations experienced by a person with a disability, may lead to an individual simply giving up trying. Despite intentions to use AT to reduce or prevent the experience of learned helplessness, it has been found that poorly designed and/or selected devices can present overwhelming learning barriers that are likely to result in feelings of failure and incompetence (King, 1999). Scherer (2005, p. 35) mentions the potential for "permanent scars on the psyche" and concludes that the key indicator of successful application of an AT should be quality of life, not task performance. Costs and benefits across social, emotional and cognitive domains need to be taken into consideration when evaluating the impact of AT (Scherer, 2005), and the dreams and goals of the individual need to be uppermost when choosing a device (Grassman, 2002).

According to King (1999), another essential human factor in AT uptake and use is cosmesis, which refers to the appearance and social acceptability of an item. This is the 'coolness' factor which may be absolutely vital in peer interactions and an influential variable in acceptance by non-disabled peers (King, 1999). Particularly with regard to AAC devices, it has been found that users and their families are likely to be concerned about the way in which communication technology will be viewed by others (King, 1999). The threat that AT use may pose to a desire to blend in has been cited as a potent reason for non-use (Craddock, 2006). The ideal AT device thus has minimal visibility and/or is aesthetically pleasing such that it is acceptable and attractive to those without disabilities (Scherer, 2005). Device obtrusiveness is a factor in problems with device acceptance (Merbler et al., 1999), because AT devices become part of the users' self-image (Cook & Hussey, 2002). If devices are seen as attractive and relevant to both user and others in a social context, they are more likely to contribute positively to the role of the individual with a disability (Grassman, 2002). For example, wheelchair users' self-image has been found to be dependent on the characteristics of the chair itself. Ultralight chairs, with their sporting con-notations, were found to impart a more positive self-image than standard bulky wheelchairs (Ragnarsson, 1990, cited in Cook & Hussey, 2002). However, with respect to the effect of voice output communication devices, there is mixed evidence regarding whether or not more sophisticated devices induce greater acceptance (King, 1999).

Grassman (2002) contends that the identity of children with disabilities is a consequence of both their understanding of their disability and their experience in social interactions. In contrast to the home environment where there may likely be a culture of acceptance, within the educational environment, segregation and limited opportunities for interaction and fulfillment of social roles may lead to identity conflict (Grassman, 2002). The addition of an AT device into the equation may result in resistance if it draws attention to the notion of a disabled self. Indeed, social embarrassment remains a common cause of AT abandonment (Scherer, 2005). Thus, for some individuals whose self-image has been able to incorporate disability, assistive devices may be viewed positively as the door to opportunity, yet for others AT may represent an unwelcome facet of their self-concept (Scherer, 2005). Thus, it's critical that AT not only foster independence and autonomy but also positive identity and enhanced self-esteem. Clearly, the use of complex technologies involves changes to notions of self and identity (Craddock, 2006).

In a recent study of post-secondary students, increased AT use was found to be associated with easier access to the curriculum, enhanced self-esteem and higher quality of life scores than those with low AT usage (Craddock, 2006). Interestingly, those that used high tech devices, such as voice recognition, screen readers and voice output, showed evidence of strong emotional attachment to their assistive devices and were found to define themselves in relation to AT. These 'power' AT users had been using technology for longer than less successful users and expressed the belief that AT allowed them to fit in, compete and, at times, out-perform their peers (Craddock, 2006, p. 22). Indeed, beyond communication, mobility and recreation,

AT is seen as an important ingredient of self-esteem for students with disabilities through its potential for increasing academic productivity (Lee & Vega, 2005). However, there is very little information available regarding the impact of AT use on an individual's emotional, personal and social goals (Scherer, 2005).

Family goals and expectations can also impact AT use and acceptance (Jeffs & Morrison, 2005; Wessels et al., 2003). The attitudes of family members are, in turn, likely to be influenced by their linguistic and cultural background (Parette & McMahon, 2002). A social climate that values technology and the psychological readiness of users and significant others will play a role in motivation for AT use (Scherer, 2005). The amount of family support may be affected by the desire for community acceptance and worry about undue attention; expectations about how quickly an individual might learn a device and how much that may improve function; the desire for interactions in natural settings; and beliefs about the likely long term benefits, especially with regard to eventual computer use (Parette & McMahon, 2002). A family centred approach, combined with cultural sensitivity, can help to promote relationships that will foster support and, therefore, positive AT outcomes (Jeffs & Morrison, 2005). Rurality is another aspect of diversity that has an identified impact on AT use. The scarcity of AT resources, services, devices and professionals in rural areas has been identified as a significant challenge to AT use (Jeffs & Morrison, 2005). Studies into the impact of diversity within the disabled population do not appear to have considered the influence of gender on AT choices and uptake (Jeffs & Morrison, 2005).

In summary, while assistive technologies have been lauded as potent and powerful allies for students with disabilities (Merbler et al., 1999), the full promise of AT often fails to be realised (Davies et al., 2001; Scherer, 2005). While AT is generally perceived as likely to be beneficial, cost barriers, lack of information, training and support, as well as individual characteristics and psychosocial factors can limit its potential. The promised rich rewards of technology can thus remain elusive amidst practical realities (Merbler et al., 1999; Wehmeyer, 1999).

Educators and AT

Despite increasing emphasis within education on the need to ensure all students who could benefit from assistive devices are able to access them, there is little consensus on how best to match students with technologies (Watts et al., 2004). The dominant model for AT decision-making is the collaborative team (Lahm & Sizemore, 2002). However, educators and other members of an AT decision-making team are likely to differ in their philosophical approach and professional priorities. Educators, with their focus on the curriculum, alongside occupational therapists, with their emphasis on skills for device use, are considered to be primary players (Lahm & Sizemore, 2002). It is considered vital that teachers play a strong role in AT decision-making to ensure that the educational goals of children with disabilities are sufficiently taken into account (Lahm & Sizemore, 2002). Moreover, regardless of the application of technology, the quality of instruction will always remain fundamental to good learning (Hasselbring & Williams-Glaser, 2000).

However, assistive technology pedagogy is a new area within education (Watts et al., 2004), and some educational barriers are so long-standing that they may not be recognized by educators (Hitchcock & Stahl, 2003). Many educators are unaware of the existence of most devices and, subsequently, are also unaware of their potential (Edyburn, 2004; Jackson, 2003). Even though educators' knowledge of AT and its applications is seen as a prerequisite for ensuring the potential rewards of its use for students (Merbler et al., 1999), it is claimed that most educators do not have sufficient technology training to be able to adequately support AT usage by students with disabilities (Edyburn, 2004; Jackson, 2003). Lack of training, even among special education teachers, is considered to be one of the main barriers to success with AT (Edyburn, 2004; Lee & Vega, 2005). This problem then compounds the already difficult challenge of appropriately matching AT devices to individual need (Jackson, 2003).

Thus, while AT is likely to be essential for inclusion and academic success for students with disabilities, its implementation often hinges on the awareness, knowledge and attitudes of teachers (Michaels & McDermott, 2003). Attitudinal barriers are considered to be a major impediment (Leung et al., 1999), and the culture of an educational institution can be highly influential in teacher attitudes (Elliot et al., 2003). There is a critical need for AT training within pre-service courses (Michaels & McDermott, 2003) as well as in-service training for experienced educators (Elliot et al., 2003). Lack of funding for training has been cited as an important obstacle in AT implementation (Leung et al., 1999). However, resistance to change is also likely to be a factor in the introduction of any technology (Elliot et al., 2003). Factors that can facilitate adaptation to change include participation in professional networks, evidence of practicality of implementation, potential for generalisability to other students, possibility of collaboration with stakeholders and a supportive school culture (Elliot et al., 2003).

Regardless, AT will likely be a key component in the IEP of students with disabilities (Ashton, 2000), and teacher comfort levels with technology in general, and AT in particular, are important predictors of educational AT use (Elliot et al., 2003). Teachers' willingness to experiment and put a device into use prior to mastery can help progress AT implementation (Merbler et al., 1999). Teachers have been found to be less inclined to facilitate AT implementation for students with learning disabilities if they need additional training, if only a small number of students are likely to benefit and if they have to alter their teaching style. Time constraints, where teachers feel they do not have time to teach themselves and/or their students, are also cited as additional barriers to AT implementation (Elliot et al., 2003). Successful AT implementation clearly depends on addressing educators' values, background and concerns as well as the needs of students with disabilities (Elliot et al., 2003; Scherer, 2005). Indeed, Edyburn (2004) claims that we are far from achieving the original vision of AT educational potential.

It is important to note, however, that current technological development often takes place at such a rapid rate that even AT professionals find it difficult to keep up-to-date (Ashton, 2000). This would seem likely to provide a particular challenge to busy educators. Given the multitude of AT devices available and the difficulties of selecting and obtaining them, one suggested solution is that teachers have access to an AT toolkit (Parette & Wojcik, 2004). Kits comprising a portable collection of tools appropriate for a particular disabled population have been suggested as an easy way to trial AT options for individual class members (Parette & Wojcik, 2004). Another recommendation, for students with mild disabilities, is the use of kits organized around learning tasks rather than individual students so that teachers could use kits to support many students in a classroom (Puckett, 2004). Such technology toolkits could allow educators to improve their awareness of available AT as well as increase their understanding of the potential of AT solutions, both of which are essential prerequisites for appropriately matching AT to student need (Parette & Wojcik, 2004). Technology can be a gateway to student, teacher, family and community empowerment (Male, 2003). However, failure to prioritise AT consideration and implementation risks depriving students with disabilities of enormous possible gains (Lee & Vega, 2005).

For many, life without technological assistance is not an option. Yet while AT can enhance access, opportunities and options, it can seldom on its own deliver quality of life (Scherer, 2005). The milieu surrounding the AT user will inevitably regulate the effectiveness of AT use and, hence, both performance outcomes (Cook & Hussey, 2002) and quality of life (Scherer, 2005). Educational environments and the culture within them will also be likely to be critical to successful AT application.

Universal Design vs. AT

The social model of disability provides an alternative framework for examining access and equity. In contrast to the medical or rehabilitation model, which focuses on deficits at the individual level, the social model proposes that environmental deficits are the cause of disadvantage. Consequently, within this model, the key to the integration of all members of society lies in accommodations at the broader societal level

(Scherer, 2005). The social model is the basis for a Universal Design, or UD, approach. Traditional design aims to cater to the 'average' individual, but advocates of UD argue that notions of an 'average' person are unhelpful (Leung et al., 1999). By contrast, the principles of Universal Design, aim to ensure that products and environments are designed to be usable by the greatest range of people without the need for adaptations or alternatives, and these principles are increasingly being applied to a wide range of projects and commercial products (Cook & Hussey, 2002). In architecture this trend is clearly seen with respect to the use of stairs and other physical barriers. It was quickly realized that designing built-in lifts and automatic doors was more efficient and cost-effective than building post-construction accommodations and, moreover, that such features were advantageous to a wide range of people, not just those with physical disabilities (Hitchcock, 2001).

Historically, AT devices have been highly specialized and, due to very small market share, very costly (Newell, 2003). They have, therefore, often been described as expensive add-ons to existing technologies (Leung et al., 1999). In fact, the use of many AT items has been compared to the retrofitting of a building—expensive, potentially isolating and ungainly (Rose, 2000). The UD approach emphasizes intervention at the environmental level. From this perspective, assistive accommodations should be an intrinsic part of all technological design such that everyday items are able to afford the broadest participation and the most flexible, individualized access possible (Rose, 2000, 2001). This provides a more efficient and cost-effective methodology than use of add-on solutions, retrofitting or later modifications (Hitchcock & Stahl, 2003; Rose, 2001). Thus, UD is recommended as far preferable to providing access solutions as an afterthought as it represents a radical and holistic approach which can address the source of access problems (Hitchcock, 2001; Newell, 2003; Rose, 2001). While AT and mainstream technology have traditionally been considered to be separate streams, recent movement toward Universal Design (also known as Design for All) has led to trends where mainstream designers aim to cater for the broadest range of abilities, or at least to ensure that disability 'hooks' that allow AT links are built into products from the earliest stages (Newell, 2003). Thus, universal design solutions are, ideally, integrated into a product or environment from the earliest stages to provide adaptability that is subtle and has the potential to benefit all (Male, 2003).

It is suggested that there are potential dangers in focusing on AT as the only technology approach for students with disabilities because to do so is to intervene at an individual rather than an environmental level (Rose, 2001). Edyburn (2004) proposes a radical reframing of AT in education towards a position more consistent with current UD trends in technology development. He argues that confusion around definitions of AT combined with systems failures to ensure sufficient AT training and knowledge amongst IEP team members give grounds for a complete rethink of the form, function and purpose of AT (Edyburn, 2004). He argues that while AT is by definition about individuals with disabilities, performance problems are clearly not specific to those with diagnosed disabilities. He therefore proposes that the notion of "technology enhanced performance" should replace the term 'assistive technology.' His rationale hinges on a belief that the potential of AT, though recognized, has not been realized and that this is in part due to the way in which educational AT has been conceptually segregated from other instructional technology.

Beyond physical access to school buildings and equipment, students require access to the materials and methods used to deliver the curriculum within classrooms and these are often inaccessible for students with disabilities (Rose, 2001). The use of learning materials such as print-based sources, uncaptioned videos, inaccessible software and undescribed graphics prohibits many students from accessing the curriculum (Hitchcock & Stahl, 2003). Excessive emphasis on assistive devices puts the onus on the learner to adapt, rather than on the curriculum to be accommodating (Hitchcock & Stahl, 2003). Universal Design for Learning (UDL) is promoted as an inclusive approach to meeting the challenge of student diversity, which can allow all students the chance to maximize their learning opportunities (Hitchcock, 2001) and address the goals of the standard curriculum (Hitchcock & Stahl, 2003). Within the classroom, UDL can be achieved through the application of "academic curbcuts" (Male, 2003, p. 20). According to Male, this should involve embracing a wide range of teaching approaches, resources and preferred learning styles so that the broadest range of learners can be accommodated. One of the key methods to achieve this is through digitalization, which provides an extremely flexible format for curricular materials that can be translated into multiple applications (Rose, 2001). Three dimensions of universal design can be afforded by the use of digital tools: (1) multiple means of representation, where individuals can use their preferred sensory mode or learning style; (2) multiple means of expression and control, which a choice of communication modes is available according to individual performance strengths; and (3) multiple means of engagement, which can simultaneously provide appropriate levels of both challenge and support (Male, 2003). For example, the same material in digital format can be read aloud by a screen reader or provided as Braille or auditory output or perhaps include hyperlinks to further explanations. The flexibility of digital formats can thus provide the same content as currently dominant print-based materials but with built-in multiple accessibility options (Rose, 2000; Stahl, 2003). Indeed, electronic documents including e-books and e-journals sourced through e-libraries are becoming the new basis for much knowledge (Boone & Higgins, 2003). This trend provides much scope for multiple access modes.

True UD should subvert the need for much AT, since many devices have been created to overcome design barriers for individuals with disabilities. Consequently, excessive emphasis on AT has been seen as contributing to a failure to embed alternative learning and access options within existing technologies (Hitchcock, 2001) and curriculum materials (Hitchcock & Stahl, 2003). Though Newell (2003) claims that true Design for All is an impossibility, a changing marketplace, catering to increasing numbers of people with a widening range of disabilities, is predicted to provide pressure to radically alter existing AT approaches. For instance, increasing rates of memory loss and dementia in ageing populations are predicted to support the rapid expansion of built-in accommodations to provide cognitive support (Hitchcock, 2001; Newell, 2003). As the user base broadens, technologies are also expected to become more free-market oriented and, thus, more responsive to user needs and correspondingly less cumbersome and 'institutional' looking. Newell (2003) argues that this shift in emphasis is also likely to improve aesthetics, functionality and affordability. Interestingly, he gives two examples of well-known products that began as AT and then moved into mainstream use—the waterbed, originally developed for rehabilitation purposes, and remote controls for domestic equipment. The uptake of such now well-known products illustrates the potential of a UD approach to benefit all. Indeed, items such as text to speech devices can act as AT but are also suitable for a broad range of mainstream students (Edyburn, 2004). Speech recognition technology is another AT increasingly making its way into mainstream use (Hitchcock & Stahl, 2003). Indeed, an increasing emphasis on UDL is predicted to cause a paradigm shift within the AT field (Hitchcock & Stahl, 2003). Nevertheless, while built-in accommodations and flexibility can better meet the needs and preferences of a wider range of users and reduce the need for much add-on AT, it is likely that many AT devices will still be necessary (Seale, 2006). It is considered essential that AT, UD and UDL are encouraged to co-exist because no one approach on its own can provide a complete solution to the spectrum of accessibility and support needs (Hitchcock & Stahl, 2003). When digital resources and flexible and portable ICT devices become more pervasive, access, participation and educational achievement for the widest diversity of learners can be realized (Hitchcock, 2001).

Overall, it appears likely that future rapid technological developments will have major implications for those with disabilities. A change of focus from AT solutions providing add-on support to a relatively small proportion of the population to consideration of broader accessibility features within mainstream ICT has the potential to address many of the most difficult AT hurdles. The simplification of the human-machine interface across newly developing technologies, such as would be afforded by increased use of voice activation, may alleviate the need for many modified input devices. The increasing sophistication, portability and co-integration of mobility, communication and information tools may also reduce the cost and awkwardness of having to address multiple access points. Such an inclusive technological approach has real potential. At the same time, however, some very real barriers including attitudinal, instructional, financial, and psychosocial will need to be addressed if that potential is to become anything close to reality.

. . .

PART 4

The Digital Divide

SOCIOECONOMIC FACTORS AND RURALITY

'Mindstorms' and 'Mindtools' Aren't Happening: Digital Streaming of Students via Socioeconomic disadvantage

ICT, equity and disadvantaged students: faith and doubt

Claims about the importance of equity in relation to Information Communication Technologies (ICTs) in education rest on the presumed benefits that access to these new technologies bring to learner outcomes and achievements within formal educational settings. If, however, there are reasons to believe that new ICTs do not provide demonstrable benefits for students in terms of educational outcomes, then learners who are *already* disadvantaged in socioeconomic, personal disability, cultural difference/marginality or gender terms will not be *further disadvantaged* with respect to educational outcomes through inequitable access to ICT resources. There would doubtless be other grounds for concern about inequitable access, but formal learning outcomes would not be among them.

New digital technologies are valuable resources for students' academic and social development. Nonetheless, it is crucial to acknowledge that simply making these new technologies physically available more equitably will not by itself lead to better outcomes for students. It may, in fact, contribute to undermining academic standards (at least the easily measured standards) for students who already face conventionally recognised disadvantages. As obvious as this may be to those familiar with the relevant debates and/or with substantial experience of actual uses of ICTs in schools, the fact remains that parents, politicians, policy makers, and many 'pedagogues' continue to cling to the strategy of putting computers in classrooms as the 'approach most likely' for redressing educational disadvantage.

Educational researchers are increasingly providing counterarguments against this quick fix mentality. A recent British survey found that although competency in using computers and associated technologies had improved, there had been little transfer or application of this learning to (other) subject competency. They also noted massive differences between the best and worst ICT implementations. Disappointment was expressed at the small number of examples found where the computer was used to extend students' creativity. As far as 'special education' was concerned, it was reported that the use of ICT was insufficient to have had palpable impact on student achievement (OFSTED, 2004).

A report from six Asian counties—Indonesia, Malaysia, Philippines, Singapore, South Korea and Thailand—details the struggle these countries (with the exception of Singapore) have faced to provide the kinds of ICT resources and teacher professional development that makes a difference to learning outcomes (*Integrating ICTs into Education: Lessons Learned,* 2004). It is worth noting that their increasing ability to provide a reasonable level of ICT resources comes at a time when these countries have high profile policies for educational change and identify ICTs as capable of realising these changes. Each of the six countries has explicit goals to move away from teacher-centred transmission style education and toward emphasising student centred learning, fostering problem-solving, thinking skills and creativity, and promoting knowledge creation and application. These are lofty goals that policy makers assume ICT will support. Yet here again the plans are based on the *potential* of ICT to better enable these changes to occur, and not on a demonstrated capacity of ICTs to do so.

Not everyone, of course, has been enthusiastic about the promise of new technologies. With the mass uptake of ICTs in the West during the 1990s, critics like Kroker and Weinstein (1994) in *Data Trash* and Shenk (1997) in *Data Smog* decried the blind acceptance of new technologies and critiqued commonly cited advantages of ICTs, including the World Wide Web. Shenk pinpointed the phenomenon of information overload, claiming that "at a certain level of input, the law of diminishing returns takes effect; the glut of information no longer adds to our quality of life, but instead begins to cultivate stress, confusion and even ignorance. Information overload threatens our ability to educate ourselves, and leaves us more vulnerable as consumers and less cohesive as a society" (p. 15). Elsewhere, Stoll (1995) observed that the Internet can contribute to a loss, rather than a gain, in productivity.

Even the term ICT (Information Communication Technology) has come under attack from leading educators like Seymour Papert (2004). The term evolved from IT or Information Technology in recognition of the emergence of communication features within computer driven technologies. Papert, however, sees this as emphasising narrow features of the technologies and of, in effect, regarding them as repositories for information and enhanced communication. For Papert, the emphasis should be on such qualities as the potential to extend the ability of humans to solve problems creatively and to serve as what Jonassen (2000) thinks of as 'mindtools.' Papert (2004) believes it is these latter aspects of school computing that are most beneficial to socioeconomic disadvantaged students.

Independent of critiques like Papert's, although resonant with them, it has become common to speak of ICTs used in school settings as 'Learning Technologies.' This, however, positions ICTs within education as an altogether different phenomenon to new technologies outside the classroom, as well as to other technologies inside and outside the classroom. This undercuts the educator's important role of enhancing connections between what happens in classrooms and what happens in the world outside classrooms, especially with students who have 'disengaged' from schooling.

It is unlikely that a single descriptor for new technologies that encompasses all its uses and potentials could be found. I would argue that it is useful to have a single term commonly accepted by educators and those outside of education systems alike. This is precisely what has happened with the term 'ICT.' Accordingly, I will use the term ICT throughout this paper, whilst acknowledging that it does not adequately describe some of the features of new technologies that are crucial to successful practice within educational settings, especially developing creativity, extending students' ability to solve problems and offering new ways of solving problems that do not exist without the new technologies.

Issues of terminology aside, a bottom line is that for more than a decade educationists and researchers have advocated using ICT extensively in classrooms based on the 'potential' of these new technologies. This potential has been promoted in very different ways. Some writers have described actual and possible practical implementations. At the other extreme, other writers have bordered on religious zeal. Numerous authors have likened the infatuation with technological development in Western societies to religious fervour. In *The Pearly Gates of Cyberspace* (1999, p. 30),

Margaret Wertheim relates the popularity of the web to a desire to live outside of the material world, eloquently describing it as "the digital re-enchantment of the world." She goes on to say, "whether or not champions of cyberspace are formal religious believers, again and again we find in their discussions of the digital domain a religious valorisation of this realm" (p. 256).

Wertheim recounts a typical 'over-the-top' endorsement of ICT by Ravitch. According to Ravitch, "we'll be able to dial up the greatest authorities and teachers on any subject, who will excite our curiosity and imaginations with wonderful video and dazzling graphs and illustrations" (p. 28). Similarly, in *The Religion of Technology* (1999), David Noble attributes the physicist Richard Seed with arguing that "God intended man to become one with God. We are going to become one with God. We are going to have almost as much power as God. Cloning and the reprogramming of DNA is the first step in becoming one with God" (vii). Likewise, some educators see ICT as a type of magic elixir, whose mere presence will remedy the problems associated with schooling in areas affected by socioeconomic disadvantage. This accounts for some large outlays in terms of hardware for 'special programs' with little thought to pedagogical considerations.

Weighing the Evidence: Different Approaches to Use

The polarised positions of fervent ICT supporters and severe critics sampled here correspond to two of the scenarios developed by Chris Bigum and Jane Kenway (1998) in their discussion of 'the ambiguous future of schooling' (see also Anderson 2000). They referred to these positions, respectively, as 'boosters' as 'doomsters.' Both camps can call on educational research evidence to support their views. Findings exist in quantitative and qualitative research alike to support either stance (Becker, 2000; Cuban, 2000; Hodas, 1996). 'Knowledge' about the effects of ICTs on educational outcomes can be described as being at best patchy. I would argue that these results reflect a fault in earlier research, whereby the type of ICT use was not considered to be an important variable and, at times, was ignored altogether. This tendency is evident in much writing on equity issues in education, where studies of equity and ICT typically involved counting computers, the overall technology budget and calculating the number of hours students spent on computers, with little thought being given to differentiating between different ways in which the technology was used (Anderson, 1993).

New evidence (Wenglinski, 1998) clearly shows that the ways computers are used has a profound effect on student outcomes. The extent is such that in one model spending more time on computers engaging in drill and practice activities leads to a deterioration of educational outcomes, whereas a different model involving high order thinking and problem solving leads to increased outcomes (Wenglinski, 1998; Jonassen, 2000). Understanding what is at stake in the differences here and acting upon this understanding is the key to promoting equity in the use of ICT in schools for groups of students with disadvantages. Indeed, it underlies the argument in this paper for ensuring *both* equitable resources for students *and* the use of models that have been identified systematically as being conducive to the scholastic and/or intellectual development of students.

During the past 10–15 years researchers have gathered some powerful evidence that ICT use in schools is associated with greater opportunities for scholastic achievement on standardised tests (Wenglinski, 1998) and enhanced opportunities to develop problem solving skills (Jonassen, 2000), provided that the technology is predominantly used in ways that involve higher order thinking and problem solving. It is important to move away from the earlier debates about whether ICT enhances learning in significant ways to new discussions about the most effective way to use these technologies in education. Some authors, notably James Gee (2003), are concerned that the ways we use ICTs and educational programs in general actually stream students from disadvantaged backgrounds into particular levels of employment. Gee (2003, p. 194) concedes that this may be 'too cynical' but the evidence gathered from a large and rigorous national American study supports his claims.

In articulating these concerns, Gee explains that

> the new global high tech economy called for lots of service workers in addition to lots of knowledge workers. The service workers needed good communication skills and a willingness to be compliant but often didn't need specialist knowledge. Thus some schools, the more advantaged ones with more economically advantaged learners, would prepare future knowledge workers via a thinking curricula while others—the less advantaged ones with less economically advantaged learners—would prepare service workers and the remaining industrial and manual workers in the new capitalism via skill and drill on the basics. (p. 193)

Is Gee's claim 'too cynical,' or is there in fact evidence that ICT use follows different models with one model leading to more service workers from disadvantaged schools and a different model producing more knowledge workers from economically advantaged schools. Wenglinski (1998) used the National Assessment of Educational Progress data collected in the U.S. in 1996 for his analysis. This data included information about students' use of ICT in schools and their academic achievement in mathematics along with other results. The participants included 6627 fourth-graders and 7146 eighth-graders. Structural equation modelling was used as one of the techniques to show whether there was a relationship between various technology use characteristics and educational achievement. Further analysis then looked at academic achievement, technology use and particular student characteristics such as race, rurality and socioeconomic disadvantage. Two very important findings in relation to ICT were as follows:

a) "Professional development and higher order thinking are both positively related to academic achievement (in relation to ICT)" (p. 29).

b) "Using computers for drill and practice, the lower order skills, is negatively related to academic achievement" (p. 29).

After providing persuasive evidence for the existence of this divide Wenglinski applies his findings to differential uses of ICTs in schools. He concludes that

minority, poor, and urban students are less likely to receive exposure to computers for higher order learning, and poor and urban students are less likely to have teachers who have received professional development on technology use. Thus, where technology matters, there are significant inequities; only where technology does not matter have these inequities been successfully erased (p. 29).

The NAEP data revealed that the combination of teachers receiving professional development in ICT and using computers for higher order tasks "were associated with more than one-third of a grade level increase" (p. 5). Unfortunately this combination was common in advantaged schools but rare in schools affected by poverty.

This large scale and carefully designed and implemented study suggests that Gee's concerns are far from mere reflexes of cynicism. Rather, they reflect what daily experience is for many students in socioeconomically disadvantaged schools. Wenglinski's study is not concerned with whether the different approaches and their associated results stem from systemic strategies to 'stream' students within economic role parameters. Whether some conscious or subconscious strategy is at work here is largely beside the point, because the effects are the same. Hence, the emphasis earlier in my argument on the importance of actively striving to apply *demonstrably successful models of ICT use* in classrooms and *not* to rely on merely counting machines and hours spent on computers as the means for pursuing and reporting equity in relation to ICTs in schools.

It is self-evident that wealthier countries and wealthier schools within countries will have higher levels and intensities of ICT resources, and that those students in wealthier countries and wealthier schools will have higher levels of home ownership than students in less wealthy countries and less wealthy schools, respectively. For this reason, investing in ICT hardware appeals to 'commonsense' as being a plausible strategic approach to addressing equity concerns on behalf of disadvantaged learners. What is not so obvious, and certainly not self-evident, is the fact that these same ICTs can be taken up pedagogically in ways that are entirely different and have qualitatively and quantitatively different learning outcomes for students. The use of statistics by education systems can mask this crucial dimension of pursuing equitable outcomes, by reducing equity to a matter of reporting increasing numbers of computers in 'disadvantaged' schools and increasing number of hours that students are engaged with computers. This is a convenient means for glossing larger problems.

Conditions of Intelligent Use

Patently, to use computers at all and, especially, to use them 'intelligently,' there are necessary levels of resourcing to be met. This explains the emphasis on 'affordability' in many critics' work. Dertrouzos (2001), for example, argues that

the rich, who can afford to buy new technologies, will use them to become increasingly more productive and the outcome is inescapable. Left to its own devices, the information revolution will increase the gap between rich and poor nations, and between rich and poor people within nations (p. 20).

While such arguments about differential levels of economic resources tell only one part of the story, levels of ownership are an integral part of the equity mix. Reports from the United States, Australia, Britain and Asia establish beyond question that schools in economically disadvantaged areas do not have the same level of ICT resources as schools in areas serving higher SES groups. Likewise, students in economically disadvantaged areas have correspondingly lower levels of home ownership of computers, Internet access, etc. Students in economically disadvantaged areas rely more heavily on the school's ICT resources than do students in more affluent areas, and as Wenglinski found with his large study in the United States, "students who use computers at home demonstrate higher levels of academic achievement" (p. 29).

The lack of home ownership in low-income groups also partly accounts for the finding of the Smerdon Report (Smerdon et al., 2000) that teachers in advantaged schools use ICT more to communicate with colleagues and parents. The Smith Family Report (McLaren & Zappala, 2002) also reported on similar disparities in Australia, where low-income families were found to have low levels of home ownership as well as low levels of access to the Internet. They cited Manktelow as being disappointed that "Australia has another great dividing range. In the age of the information economy, modems—not mountains separate the population." In Canada, Reddick et al. (2000) found that by 1999 about two-thirds of upper income families had home access compared to one-quarter of low income families. In Asia, even greater disparities can be found, although reliable statistics are harder to come by. Home ownership of computers in Hong Kong can be sorted according to the type of housing status. For example, in villages the level was 2.2%, housing schemes 14.3%, public housing 31.7% and private housing (50%) (Ure & Kuen, 1996). In Thailand the government has been concerned about lack of home computer ownership in less affluent homes and have introduced the 'ICT PC Project' whereby one million low cost PCs have been sold to the general public at a discounted price made possible by government subsidy. In Singapore, families from low-income groups have been invited to attend ICT training sessions, and many of these families have been chosen to receive free PCs if they attend the sessions and show achievement.

This obvious disparity in ownership of ICT resources means that many students living in areas affected by socioeconomic disadvantage have only one 'shot' at using ICT to enhance their educational development, and that 'shot' is at school. If the schools in these areas adopt less beneficial—if not positively *harmful*—models of ICT use, as evinced in Wenglinski's large study, this amounts to ICTs being used as tools that actually widen the gap between 'haves' and 'have-nots.' Given that powerful models for effective use of ICTs have been advanced by educators like Seymour Papert since the time desktop first became available in the 1960s, it is imperative that teachers, schools and education systems ensure that such models are adopted.

Seymour Papert and Powerful Computing

More than twenty years ago Papert (1980) established that the main benefit of computing in education lay in its capacity to stimulate the mind—albeit only if ICTs are used in particular ways. His bold prediction that one-to-one computing would become common by the present day was derided. Yet his most insightful observation was that "there is a world of difference between what computers will do and what society will choose to do with them. Society has many ways to resist fundamental and threatening change" (p. 5). Papert's fears that computers would not be used in ways that engaged students at high intellectual levels have been realised.

Papert likewise envisaged computers as powerful media for enabling cultural change and for addressing issues and factors of isolation and other forms of social disadvantage. Computers, said Papert (1980, p. 4), "can be carriers of powerful ideas and of seeds of cultural change [and] can help people form new relationships with knowledge that cuts across traditional lines." With others at MIT he designed the LOGO computer language as a new means for students to take control of computers, rather than being controlled by them—as they were with drill and practice or tutorial software. He saw advantages in particular uses of ICT, such as programming, that supported activity based learning and not only involved higher order thinking and problem solving but also led to increased opportunities for students to develop metacognitive processes. Papert described components of LOGO like the turtle, as 'objects to think with.' He argued that within the objects and their functions there was "an intersection of cultural presence, embedded knowledge, and the possibility for personal identification" (p. 11).

LOGO is still used around the world in many elementary and secondary schools. It has evolved to be more colourful and graphically stimulating in such versions as Microworlds. While LOGO has undoubtedly had some impact within formal education, its use is still marginal by any measure. It is, of course, utterly dwarfed by the massive use of drill and practice and tutorial software, particularly in schools serving economically impoverished communities.

More recently, David Jonassen (1996) has applied many of Papert's ideas to other software applications. Programming is not an area many educators are keen to engage with. Consequently, applying the general ideas proposed by Papert provides one useful strategy for trying to change the culture of classrooms. In a manner reminiscent of Papert in 1980, Jonassen (1996) begins his book *Mindtools* by lamenting that ICT is not typically used in classrooms in ways that lead to intellectual engagement, and proposes a positive alternative: namely,

I recommend a significant departure from traditional approaches to using computers in schools. I promote the idea of using selected computer applications as cognitive tools (which I call mindtools) for engaging multiple forms of thinking in learners (p. 3).

He defines mindtools as "computer-based tools and learning environments that have been adapted or developed to function as intellectual partners with the learner in order to engage and facilitate critical thinking and higher order learning" (p. 9). Throughout his book Jonassen offers practical examples of how database, spreadsheets, concept mapping, expert systems, microworlds, visualisation tools, multimedia and computer mediated communication tools can be used as 'mindtools.' He emphasises the ways some ICT applications can be used for 'productivity' tasks, like tying a set of text or creating a spreadsheet. He differentiates such uses from 'mindtools,' however. For Jonassen, mindtools are used in ways that lead to higher order thinking. He argues that these tasks should be 'authentic' and that "learning tasks are situated in some meaningful real-world task or simulated in some case-based or problem based learning" (p. 11).

Notwithstanding problems involved with descriptors like 'authentic' and 'real-world,' Jonassen's underlying principle seems to me to be sound. Many school based tasks could be described as 'authentic' within their own context. Likewise, 'real-world' implies that the classroom or school is part of some imaginary world. When students go from teacher preparation into classrooms, they often describe the classroom as 'the real world' in comparison to the presumably 'unreal' world of tertiary education. Some workplaces and situations outside of schools have particular ways that make sense only in that particular environment and are just as 'real' or 'unreal' as classroom contexts. Classroom activities are often more effective as contexts for learning when they are linked to aspects of students' lives beyond the classroom and build productively on learners' needs, prior experience and interests. An even better proposition might involve ICT activities embedded in tasks considered to be important by the community and carried out in community spaces (Bigum, 2002, 2003; Lankshear, Synder & Green 2000). According to Lankshear, Synder and Green, principals should provide incentives for teachers "to develop effective links with other schools and community organisations, with a view to sharing expertise and resources [and] undertaking collaborative projects involving integration of new technologies' (p. 158). Carroll (2000, p. 116) sees an opportunity for schools to become 'nodes' of a broader network and maintains that "we must recognise that schools and classrooms are becoming nodes in networked learning communities. We must begin to think about how to organise learning in networked communities and not limit learning within the boundaries of the classroom and school buildings." Gee (2003, p. 192) speaks of 'affinity groups' where members "bond to each other primarily through a common endeavour." It is up to principals, teachers and students to explore 'common endeavours' that will bring communities and the world outside the classroom closer to classrooms or to bring the 'classroom' to the community.

ICTs and poverty: Issues and response

Issues of socioeconomic disadvantage and ICT tend to dominate the community debates along the lines of the 'digital divide' discussions in the United States, where statistics have been used to demonstrate the differential between the 'have' and 'have nots' in terms of home ownership of equipment and access to the Internet. Van Dijk and Hacker (2003) point out that the gap in home ownership of ICT resources based on income was acute during the 1980s and 1990s but has now closed to some extent. However, they cite evidence that the gap in *skills* and *use* are increasing. A task force set up by the Schools Commission (cited in Connell & White, 1991) identified five dimensions to poverty. These included:

»   Inadequate family or community income;
»   Vulnerability to the changes in the labour market and economic depen-
    dence;
»   Lack of organisational power and being excluded from collective resources;
»   Damaging social and physical environments;
»   Cultural marginalisation. It is also important to remember that poverty is
    more than merely a lack of money and that other factors come into play apart
    from merely income and what ICT resources can be afforded.

Dealing with poverty demands a close examination of organisational structures that may serve to further disadvantage some children and advantage others. Connell (1993) claims that the total 'social investment' in the education of economically advantaged children far outweighs the effect of any compensatory funds directed to poor children. To maximise the effect of funds directed to ICT in disadvantaged schools, it is necessary to ensure that benefit is gained by target groups and that whole school change is encouraged and effective models of ICT are employed. Unless these conditions are met, the use of ICTs in many schools serving disadvantaged populations will amount to just one further dimension of poverty—a condition that helps maintain the vulnerability of already disadvantaged students to changes in workplaces, and that marginalizes them from the qualitatively different ways that advantaged groups use ICTs.

A small scale example of using ICTs along lines captured in Papert's and Jonassen's ideas of 'mindstorms' and 'mindtools' that had measurable positive impact on learning outcomes of disadvantaged learners can be provided, based on work by this author (Anderson, 2001). An initial pilot study (Anderson, 1999) involving one student with intellectual disabilities assessed in the 'mild to moderate' range suggested that using ICTs in accordance with constructivist concepts and principles could overcome impediments to academic success that had previously seemed insurmountable. The approach and techniques used in the pilot study were also subsequently used successfully with a larger group of similar students who had experienced long term failure despite extra 'expert' remedial help. Mirroring the work of Papert, Jonassen and others, this program stressed the importance of engaging students in higher order intellectual engagement notwithstanding their formally assessed 'mild to moderate' intellectual disabilities.

The outcomes of this project illuminate the larger argument in this paper in quite powerful ways. All participants had become disengaged with formal schooling, had experienced long term failure despite additional one-to-one efforts of remedial teachers, had not developed even basic reading skills, and had built up diverse 'blocks' to school based learning. In order to measure any potential effects of the intervention, rigorous data collection of student outcomes associated with the development of verbal communication skills and reading skills and the development of new ICT literacies was built into the design. Verbal communication skill development was assessed by videotaping peer tutoring sessions (where the participants were the tutors) at regular intervals during the intervention and recording and comparing performance on seven key indicators. Reading performance was monitored by the use of two standardized reading tests administered across the sites by the same qualified guidance officer (pre- and post-testing) and developing ICT literacies were assessed by the collection and review of portfolios (Anderson, 2001).

Within entire Australia-wide cohorts of students with equivalent measures of learning disadvantage to those participating in the intervention study, only very rarely do individuals manage to attain formal measured reading ages by the end of elementary school. By contrast, all of the students involved in this intervention study based squarely on a 'mindstorm' and 'mindtools' pedagogy achieved formal reading ages and became engaged in learning in a manner that was sustained during their final elementary grade levels and into secondary school. Significant wider development of verbal communication skills and computer skills was demonstrated by all participants in the study. This group of students had effectively been marginalised by conventional practices of dealing with students with intellectual disabilities by over-emphasizing skill based development in the basics while ignoring the development of creativity, problem solving, higher order thinking and metacognition. This parallels the ways that many students impacted by poverty are treated.

Advocates of ICT use in schools have often emphasised the potential of ICT for improving curriculum offerings across the board by integrating new technologies across the curriculum, and thereby providing inspiration to teachers and motivation to students. Despite our best intentions, however, if we fail to employ effective models of ICT use we may actually end up maintaining students' current disadvantages—if not *exacerbating* them—rather than contributing to enhanced learning outcomes.

The Digital Divide, Poverty Alleviation and ICT: Some Illuminating Case Studies

The Internet is still an alien construct to the majority of people in the world. The gulf that separates this majority from those who enjoy access to computing technology is referred to as the 'digital divide.' The term is simple and implies a dichotomy, hence early efforts to first describe and subsequently to close the digital divide centred on who had access to technology and who did not. A logical 'follow on' was to ensure that as many people as possible had access to hardware and software (van Dijk, 2006), thereby suggesting that solutions are digital. However, exploration of the contours of the digital divide only serves to highlight its intricate and undulating topography and its inherent complexity. In his review of a book by Mark Warschauer ("Technology and social inclusion: Rethinking the digital divide"), Gray (2004) noted that a 260 page limit rendered analysis rather thin in view of the fact that the book probed "social relationships, communities, and institutional structures" (p. 294). This would in fact seem to be fairly widespread in literature regarding the digital divide, as investigations into it must inexorably highlight the cultural, sociological, political and economic realities that impact on it.

Gray's (2004) observation is consistent with those of van Dijk (2006). According to van Dijk, the digital divide's implied dichotomy is the source of misleading and deep ambiguity for four reasons: (1) it suggests there is a clear gap between two recognisably divided groups; (2) it indicates that bridging the gap is problematic; (3) the groups are defined in absolute terms (those who 'have' and those who 'have not') and (4) the divide is fixed and unchanging (static). van Dijk observes that the bias is normative as well as technological and is dynamic in nature. In writing of the African experience, Fuchs and Horak (2008) are supportive of this thesis, but are critical of many solutions advanced by scholars because such interventions do not address deep seated and complex societal issues.

As far as is possible within the parameters of one chapter, the purpose of the current review is to draw a 'mud map' of the digital divide. This will be done by casting its main characteristics in broad strokes. In the interests of illustrating the practical benefits that can be enjoyed by people whose position lays beyond the digital divide, the chapter will conclude with the presentation of four case studies. These will highlight the practical benefits that can be extended to people in remote and rural communities by Internet provision.

Characteristics of the Digital Divide

Economic and social contours of the divide in OECD countries

According to the Organisation for Economic Co-operation and Development (OECD) (2001), in 2000, 90% of all Internet users in the world came from its 30 member countries. Given that the United Nations (UN) (2006) has listed 192 nation member states, this leaves 10% of Internet users in the world coming from the remaining 162 countries in the world. Therefore the first identifiable digital divide contour is an economic one. The advent of the Internet has provided for rapid growth of competition and business expansion in OECD countries, stimulated new investment and innovation and "liberalisation of telecommunication markets" (OECD, 2001) and consequently made significant contributions to productivity and profitability. This was achieved through the introduction of competitive broadband in many OECD member countries, with attending exponentially increased rates of Internet take-up and use. According to Ono and Zavodny (2007), who surveyed patterns and determinants of Internet use in four OECD countries: the U.S., Sweden, Japan, South Korea and one non-OECD, but nevertheless similarly wealthy, country Singapore, there were some individual differences between countries. However, the common economic commitment of each of these countries to growth and productivity enable a 'first brush' examination of features of the digital divide.

English Language: Most Internet websites use the English language or Chinese with English dominating important sources of information. This could explain the comparatively slow take-up of the Internet in countries where English ability is low, such as Japan (Ono & Zavodny, 2007). Since English is also the language of e-commerce (OECD, 2001) this would perhaps explain the proliferation of English language classes in South Korea (Bender, 2005). Singapore uses English as the main language in the sectors of education, health, government and commerce, and in regard to Sweden, the other non-English speaking country examined by Ono and Zavodny (2007), a high percentage of its citizens speak English as a second language.

Improvements in Broadband Internet and their 'flow on' effects: The government of South Korea set about to build a 'Knowledge Based Society,' and it emphasized the potential of the Internet in advancing education of citizens and played an active role in the construction of a high speed broadband network (Picot & Wernick, 2007, Youngbae, Jeon & Bae, 2008). Consequently according to Youngbae et al., South

"Korea has been leading the world in broadband Internet access services since 1999" (p. 307). By contrast, Ono and Zavodny observed that an authoritarian approach adopted by Singapore's government 'kick-started' Internet information structure but restricted content, thereby possibly inhibiting its potential growth.

The introduction of broadband Internet has given consumers an 'always on' service, and in most cases they pay a monthly tariff rather than for the amount of time used (van Dijk, 2006). As the quality of the broadband improves, so the quality and enrichment of content improves exponentially, further increasing the capacities of the Internet to satisfy the needs of people seeking more than communication, information and business communication (Goldfarb & Prince, 2008, Hitt & Tamb, 2007, Ono & Zavodny, 2007, van Dijk, 2006). Furthermore, as competition between providers drives the costs to consumers down, more people find that they are in a position where they can afford to be connected to the Internet (OECD, 2001). Thus improvement in quality and lower cost have meant that within OECD countries, more less affluent members of these communities are taking up Internet subscriptions. In regard to the adoption of broadband Internet by people in the U.S., Goldfarb and Prince concluded that:

While income and education positively correlate with adoption, they negatively correlate with hours spent online. Given our results, we argue that the most likely explanation for this finding is that low-income individuals spend more time online due to their lower opportunity costs of leisure time. In particular, the pricing structure of the Internet, with both fixed connection and near-zero usage fees, leads to a negative correlation between income and time online among those who have connected. We interpret the fact that low-income people are particularly likely to do time-consuming, inexpensive activities online as support for the role of the opportunity cost of leisure time (Goldfarb & Prince, p. 14).

Prieger and Hu (2008), however, noted that in the U.S. there is also a racial flavour to the roll out of broadband technology. While black and Hispanic citizens spend more time in activities that by definition require high quality broadband services, such as downloading music and online gaming, there is a gap in service quality to such groups, compared to that enjoyed by white citizens.

The Urban/Rural Dimension: According to Wood (2008) the competition factor compounds the disadvantage suffered by rural citizens because the pursuit of profits mean that large companies prefer to focus on areas with higher population density. Wood, in discussing the American experience, observed that these companies favour metropolitan markets and are actively trying to shed their rural landlines, leaving rural people with outdated infrastructure. This was supported by Warren (2007) who described the predicament faced by rural people as a 'digital vicious cycle.' He contended that rural people stood on the disadvantage side of a widening gap as their metropolitan counterparts enjoyed the fruits of broadband competition. Warren's term for this disadvantage was 'disbenefit.' Due to the rising utility of the Internet, those who are excluded from the opportunity to use it are subject to social exclusion and increasing disadvantage (see also La Rose, Gregg, Strover, Straubhaar & Carpenter, 2007).

The Poverty Dimension: Gilbert, Masucci, Homko and Bove (2008) accessed perspectives of low income African American women living in inner cities in relation to using 'telemedicine,' or health care disseminated through the Internet from a central institution. Gilbert et al. found that the women overcame their initial inefficacy through the mediation of shared networks, and that libraries or other community centres were essential for providing access and the combination of ICT training with a necessary service. Therefore Gilbert et al. suggested that investigation of the digital divide must incorporate the perspectives and experiences of those who access and use the resources, rather than focus on disparities in access to computers and the Internet. This observation, drawn from people living in an influential OECD country but nevertheless suffering major economic disadvantage, provides an important link to those who might well fall through the 'World Wide Web.' These are citizens of the developing and undeveloped countries of the Third World.

Economic and social contours of the digital divide in developing countries

Because of the wide range of countries involved it is impossible to mirror the previous section where defining aspects of the digital divide within OECD countries were outlined. However, in keeping with the observations of Gilbert et al. (2008), where it was recorded that perspectives of poor people who had to use Internet technology within a particular framework could inform dialogue in regard to the digital divide, it is noted that because of the enormity of preexisting problems intervention in developing countries must of necessity be piecemeal in nature. Many of the problems outlined in the previous section also occur within developing countries. However, profound preexisting problems render these even more complex. For example, Mathur and Ambani (2005) noted that within India, which stands at the very centre of technological advance in the Third World, 3/1000 people own computers and 66% of India's people have not benefited from ICT expansion. They observed further, "in a country of 1 billion people, millions of Indians are connected to the Internet, but millions more are not yet even connected to electricity" (p. 346).

It could be that this sort of stark reality of grinding inequity has led researchers such as Fuchs and Horak (2008) to suggest that the solution lies with fundamental changes in society, thereby taking the discussion well beyond what can be done at a practical level into the realms of philosophy and international politics. Therefore it is observed that, when it comes to developing nations, any sustained examination of the term 'digital divide' provides a microscopic and therefore perhaps overdetailed view of world societal divisions that are long standing and ineffaceably complicated. It could well have been this aspect that led van Dijk (2006) to conclude that the digital divide is a "container concept carrying too many meanings" (p. 222) and to call for more refined and interdisciplinary development of theory, more precise conceptual definition, and both qualitative and longitudinal research.

Beyond the Digital Divide

The current analysis therefore does not seek to underestimate the enormity of problems facing communities in developing countries or to overestimate the impact of the interventions under discussion. The four case studies presented in this paper outline practical benefits to the communities under discussion such as increased profits for crops, better farming techniques, opportunities for e-commerce and the production of wealth, as well as new forms of entertainment. The rationale behind such an approach is one of 'thinking locally but acting globally.' The vexing question that will remain is whether it is really possible to retain 'local thinking' in the face of the ever increasing influence of the global media. This leads inevitably to consideration of what is an ethical approach to the provision of ICT based community centres located deep within areas of grinding poverty in the Third World.

Poverty alleviation

A search of the Web of Science/ISI database revealed 288 papers under the search term 'poverty alleviation.' The addition of 'technology' reduced the field to 19. A further search under 'poverty alleviation and information communication technologies' yielded two papers. Despite this dearth of empirical research associated with ICT and poverty alleviation, Flor (2001) claimed that ICT and poverty are indisputably linked. Flor argued that "since the 'information society' concept was introduced in the seventies, the correlation between access to information and poverty has been widely acknowledged" (p. 1). Mathur and Ambani (2005) pointed out that "the application of ICT solutions for the development of rural India and other developing countries opens up a vast range of possibilities" (p. 345). The example they provided was of a project in India where because of judicious installation of rural cybercafés using a mixture of dial-up and wireless technology, rural people could be relieved of the burden of having to travel for personal business. Therefore they achieved better prices for their produce and were able to apply better management practices for their farms. This sort of approach is consistent with the reflections of Gilbert et al. (2008), who noted that Internet efficacy for poor women came through publicly available technology in local community centres or libraries where assistance was freely available within a context that was meaningful for them.

In the case of India, where many people rely on micro finance to reduce, mitigate and cope with risk, Mathur and Ambani (2005) noted that the cybercafes had also provided poor people with the capacity to conduct their affairs in a way that enhanced their efficacy with the technology in an atmosphere of support. The growing importance of ICT as a basic need of society was highlighted by van Dijk and Hacker (2003), who claimed that "in information and network society, relative differences in getting information and lines of communication become decisive for one's position in society, more than in every other society in history before" (p. 324).

There is little doubt that people in developing countries of the world suffer levels of illiteracy and disadvantage that are beyond the imagination of many in more developed societies. Poverty has been defined by the Asian Development Bank as "the deprivation of essential assets and opportunities to which every human is entitled" (Flor, 2001, p. 5). These essential assets and opportunities refer to many aspects of modern life involving ICT, including access to a basic education which would ideally be enhanced using ICT. Basic communication, access to information on farming techniques, basic health services and information are all linked to the basic provision of ICT. According to Drori and Jang (2003, p. 156), "cross-cultural differences in IT resources and capabilities are sharpening global inequities and, moreover, setting a new geography of global centrality/marginality." In a similar vein Frieden (2005) suggested that ICT had the capacity to "prime the pump" (p. 596) of national economies, in that consequent enhancements to productivity would inevitably raise living standards.

It is in the interest of providing some detail as to how the 'pump can be primed' (Frieden, 2005) that the following four small scale case studies are provided. The case studies are small and local in nature because they have emerged in response to local needs. These four projects have, in diverse ways, contributed to the alleviation of some aspects of poverty by fulfilling both the economic requirements of long term sustainable development and by providing opportunities for people to contribute to their communities, thereby fulfilling an important social purpose (Johnson, 2005).

The Jhai Project in Laos

Laos, a country in southeastern Asia is northeast of Thailand and west of Vietnam and has a population of approximately six million people. Lao is the official language and other languages spoken include French, English and various ethnic languages. It is a country with a very poorly developed infrastructure. For example, it has no rail system and an elementary and poorly maintained road system. Telecommunication systems offer little external or internal access, and electricity is restricted to a few areas with higher population densities. A large percentage of the population (80%) is engaged in subsistence farming. The project discussed in this paper is taking place in the Hin Heup district of the Vientiane province, which is approximately 100 kilometres from the capital, Vientiane. The nearest markets are situated in Phong Hong, which is approximately 30 kilometres away. Facilities for transport are scarce and torrential rain in the wet season makes the roads almost impassable.

Technical features of the project

Lee Felsenstein, developer of the Osbourne computer and pioneer in the development of publicly available microcomputers, designed a compact, rugged computer with the power of a pre-Pentium machine specifically for the Jhai Foundation project and the conditions in Laos (http://www.jhai.org). The machine has no moving parts, is small and compact, and has a waterproof case designed to counter the onslaught of the southeast Asian monsoon season. The main processor is a 486-type chip, which allows the use of a heat sink rather than a fan, thereby eliminating the common problem of fans seizing up in adverse conditions. More powerful processors, like recent Pentiums and AMD chips, require large fans. The Apple G5 requires multiple fans as well as a liquid cooling system. These are all potential problem sources in harsh conditions. Importantly, whereas a standard desktop computer requires 90 watts of power, the project machine draws only 12 watts. The machine uses a small energy saving LCD screen, and flash memory chips have replaced a conventional hard disk drive. Finally, the machine has been designed to withstand formidable conditions of different kinds for a minimum of 10 years with little or no maintenance. These conditions include torrential rain, choking dust, and intense heat and humidity at different times of the year. The machine will ultimately be capable of mass production for less than US $400 per unit.

On the basis of an initial investigation solar power was rejected as a means of powering the unit, due to the cost involved as well as the adverse cloudy and gloomy conditions during the wet season. Pedal power subsequently proved to be a very sustainable method of providing power, and at just one-third the cost of solar energy. A prototype pedal generator linked to a standard automotive battery produces five minutes of computer use from each minute of pedaling.

Wireless connectivity is provided by a standard 802.11b card linked to an antenna located on the thatched roof of the bamboo structure in the Phon Kham trial site. This is in turn linked—via a solar powered repeater station in the hills—to a local Internet Service Provider (ISP) in a larger centre. The system relies on standard and relatively inexpensive wireless hardware components.

Perhaps the most interesting component in the overall system is the locally adapted Linux operating system and software for word processing and simple spreadsheeting. Dravis (2003) describes the system as a local version of KDE, known as LaoNux on the Debian Linux distribution. Technical work on this component of the project was conducted by Anousak Souphavank, a former IBM programmer now working at the National University of Vientiane, with students and lecturers who provided voluntary assistance. This complex technical challenge involved creating a custom Unicode to support the Laotian font set. The result is a stable version of Linux using the local Laotian language.

Crucial software components include a web browser for navigating the Internet, a local language version of an open source word processor and spreadsheet package, and a Voice over IP (VoIP) system that makes local and international telephony possible through the standard phone system. These software applications meet the needs and uses of the system expressed by local people. Thorn (2004) identified some of the planned uses. According to Thorn,

> Right now, the villagers have no way of telling what the market is in the big towns they sell their stuff to, telling what the weather report is for their crops, things like that. This will absolutely change that. Plus, they will be able to talk to relatives in America some of them they haven't seen in decades (n.p.).

The hopes for utilising the system have been realised in the trial village. Plans are currently underway to extend the program to four other Laotian villages and to sites in up to six other countries. In this project the community expressed a need for ICT, rather than having it imposed on them, and in addition the uses of the community hub were decided at a grassroots level. Starting in a small way meant that sustainability was more likely and the community would be able to build on the advantages afforded by the technology (Anderson, 2005).

Ethical concerns

The obvious advantages of the project include increased national and international communication using VoIP, since prior to the introduction of the technology community members had no access to telephone communication. Opportunities to increase wealth through better market information and farming practices are evident in the development of the community centre. Conversely, negative effects may arise from a shift from mainly subsistence farming with a minor amount of produce sale to a more Westernized commercial approach to agriculture. In this model, the organic techniques seem more in tune with the context, but traditionally the village culture is embedded in a subsistence way of life. Changing this model will have unpredictable effects. Exposure of the village people to the Internet may or may not have a significant effect. Singh, Zhao and Hu (2005) and Corbitt, Peszynski, Inthanond, Hill and Thanasankit (2004) produced research to confirm that the Internet is not culturally neutral and that empirical examination of sites reveals embedded code that contains culturally specific notions of such attitudes to gender and power relations in society. Straub, Loch and Hill (2001), however, demonstrated that these attitudes are not necessarily automatically assimilated by users of the web. In fact, their research demonstrated a 'cultural resistance' to these foreign attitudes, further reinforcing the unpredictable nature of the introduction of ICT.

The LINCOS Project

The LINCOS program was initiated in 1998 as a joint initiative between the Costa Rican Foundation for Sustainable Development, the Media Lab at the Massachusetts Institute of Technology (MIT) and the Costa Rica Institute of Technology. The major aim of LINCOS is to promote sustainable development opportunities for communities, mainly in isolated and rural communities, through increasing the access of ICT. This model involves a transportable community hub with more expensive infrastructure and support requirements than the Laotian model outlined above. Sheats (2000, p. 41) described the LINCOS model for the building infrastructure as follows:

> [Physically], the fundamental basis of the LINCOS digital town centre is a standard ISO shipping container (2.4 x 2.4 x 6.1 m) remodelled and equipped with a set of IT and wireless communication equipment. It is Internet-linked via satellite, with standalone power source and measurement capabilities for medical and analytical applications.

This LINCOS Project, based in Costa Rica, has been in operation longer than the Jhai Foundation initiative, and offers some features that could be taken up by the Jhai Project. For example, the LINCOS model uses low cost probes and testing equipment to measure important soil characteristics. These measurements are sent via the web to distant scientific laboratories that provide advice on deficiencies in the soil or requirements of particular crops. In the case of the organic production approach used by the Laotians, this would mean adding organic phosphorous, nitrogen, or whatever other nutrients were required to ensure higher crop productivity or quality. Another interesting application is the training of a health monitor who uses the Internet, software and basic instruments to provide health care and preventative advice. LINCOS, however, uses Microsoft Windows operating systems and software. It is a considerably more expensive option and possibly not suited to the types of locations potentially served by the Jhai model. Some of the same community uses could, however, be supported.

The LINCOS model has developed methods of using ICT for educational purposes. It has adopted principles outlined by Papert and Cavillo (2001) for applying constructivist learning principles in practice. Papert has had direct involvement in the project through LINCOS's partnership with MIT. In Papert's model, students use the technology to solve genuine community problems, such as building bridges and roads, using word processors to record project proposals, spreadsheets to facilitate genuine budgets and various other ways of solving social problems. They also take control of the machines by engaging in basic programming, using Logo, and even engage in linking computers to other devices like simple robots. Using a model involving a preconceived and transportable building could be interpreted as an 'imposition' rather than as a response to a community need, although it has the advantage of being completely set up for its intended purpose. Importantly, the coordinators of the project negotiate with the community on how the facility in used is various locations, thereby increasing the idea of 'community ownership' of the hub.

Ethical concerns

Papert laments the simplistic way that ICT has been used in Western schools as opposed to his vision of a more constructivist, problem-solving model and sees an opportunity to introduce these techniques in different cultural settings with perhaps less baggage and resistance to change. But how can he and the project workers predict the outcomes of the introduction of a way of thinking that may or may not comple-ment the existing local culture? As in the Lao model, new ways of farming have been introduced based on Western agricultural techniques and therefore paving the way for future changes with unpredictable results. Similarly, the introduction of medical models based on the West seem beneficial at first glance, but the future of this may be uncertain given the transportable nature of the units. What are the results of introducing a community to satellite based Internet coupled with computers and ICT peripherals and then moving this equipment to another location?

Mobile phones in Bangladesh

An example from Bangladesh of an elegant but simple model with far reaching positive consequences was outlined by Flor (2001) and was initiated by the Grameen Bank of Bangladesh. Flor (p. 7) records that the project

> Would put a mobile phone in some 45,000 villages, giving residents access to ICT. Each mobile phone is acquired by an individual through a small loan from the bank. This phone becomes a village telephone service provider, earning income for the owner besides provid-ing a much-needed utility to the community.

Of course many communities in developing countries do not live in areas serviced by mobile phone networks, but for those areas covered by such services, this project is ideally suited. Mobile phones are increasingly equipped with the types of functions formally associated with full computer systems such as the capacity to connect to the Internet, and therefore the significance of this idea is immense. Flor reports that the mobile phones are already being used utilised in a basic form of e-commerce, whereby farmers use the mobile devices to check out the latest market prices.

The Torres Strait Islands Project

The Torres Strait Islands consist of Thursday Island, Horn Island and Hammond Island (as the main administrative centres) and 16 outer islands stretching from north of Queensland, Australia, to within sight of the Papua—New Guinea coastline. These islands have a total population of approximately 8000 people with a background of predominantly Torres Strait Islander or Aboriginal culture. Although the islands are remote (at the time of the project planning), they had reasonable access to ICT in the form of infrastructure that supported the access to the Internet via dial modem at a reasonable cost through the Australian telecommunication carrier Telstra. Some computer use was evident in the local schools, although little evidence was found to demonstrate a reasonable rate of community access through home computers. Most island communities had access to a limited range of radio and television services, although some smaller, inhabited islands remained without access to basic services. Where services were available they were provided by digital microwave radio to Digital Radio Concentrator Services on particular islands.

In comparison to some remote centres, many of the islands were reasonably serviced by the available infrastructure. However, it was apparent that community uses of ICT were low and availability of training to facilitate effective use of new digital technologies was virtually nonexistent. In response to this, the author and staff from Queensland Health (State Health Authority) developed a submission, in consultation with the local people, to the 'Networking the Nation' grant scheme initiated by the Australian federal government. Underpinning the initial planning for the successful grant of $5.5 million was the need to further improve the communication infrastructure to support higher speed Internet links to support the development of community hubs. These hubs needed to support features identified by the local communities as important means of overcoming the problems associated with living in very remote, isolated areas, and importantly, the facilities needed to be readily accessible by the general community and designed to meet their needs.

An important need outlined by community members and community councils involved the lack of medical services, particularly specialist medical services. Many older people felt extremely reluctant or threatened by the prospect of traveling long distances by boat or plane or simply did not want to leave their traditional island land for any purpose—even to seek much needed medical assistance from specialist doctors. New digital technologies planned for the community hubs included facilities to enable videoconferencing links, so that the visiting or resident medical practitioner and the patient could link through to the city to engage in telemedicine sessions. In a similar vein, community members could appear in distant courts via the facilities as witnesses or defendants via the teleconferencing facilities. Although face-to-face meetings of tribal leaders were important cultural events, some meetings could take place through teleconferencing via a 'bridge' system to link up multiple participants in multiple locations. In many ways the use of videoconferencing could assist to alleviate the tyranny of distance in accessing important services, including the essential element needed for success in accessing and using digital technologies—training and technical assistance.

Although the physical design and position of the community hubs were to be negotiated with the local people, a general agreed set of resources was developed and these resources included plans for a model based on a building equipped with a main server and thin client connection to computer terminals that would provide high speed access to the Internet and a set of commonly used programs such as word processors, database, presentation software and spreadsheets. Access to the computers would be through a 'smart card' system, thereby giving affordable, but not free, access, thereby strengthening sustainability.

Although major uses of the 'hub' such as video-conferencing for telemedicine, law and education had been negotiated across multiple communities, each community could develop uses that were seen as valued or important in their particular context. For example, one island community (Moa Island) had contracted the author to work with them on developing multimedia skills so that children engaged with learning at the local school could have access to more materials that linked to the curriculum but at the same time included appropriate cultural content. Through the use of digital photography and moviemaking, along with text production, specific curriculum materials were developed to highlight local culture. Many of the local people had the view that curriculum materials reflected the culture and context of the Australian mainland rather than their own culture.

An important consideration in the project was that the local culture varied significantly from island to island; therefore a one-size-fits-all solution was not feasible. Another example was an island community with a strong focus on Indigenous artwork wanting to use the hub to set up an e-commerce centre to enable international marketing of their products. Other island communities sought to extend the use of the smart card to the community store and other outlets for transactions, so that communities would effectively become 'cashless societies.' Each individual's funds would be transferred to the smart card and used to purchase all required goods and services, thereby eliminating the need for currency. In fact, on some of the smaller islands, currency loss becomes a serious issue as people leave and take currency with them, leaving the community short of available coins and notes.

Ethical concerns

Some superficial effects of global media were obvious to the researcher in the form of the Torres Strait Islander youth wearing U.S. basketball singlets and paraphernalia and incorporating the slang of American rap culture in their everyday speech. Traditional ways of life in this area include hunting. Laws respected by the Australian government allow the hunting of (normally protected) turtles and dugongs, but this coupled with the use of modern boats, outboard motors and high tech devices leads to a worrying amount of over-fishing of those species. Opening up the sale of traditional art on a global basis through e-commerce could have far-reaching and unpredictable consequences. As far a recreation is concerned, the youth, in a similar way to Australian mainland culture, particularly the young males, are attracted to violent video games. Researchers currently cannot agree about the effects of this exposure in urban settings let alone in traditional settings in the Torres Strait.

Conclusion

It is clear from these ICT projects that grassroots initiatives can make a difference to communities and assist in poverty alleviation. The 1997 Human Development Report outlined six essential actions to alleviate poverty, the first recommendation stated that: "Efforts should be made to develop viable ICT poverty alleviation programs. These programs should be coordinated across agencies in the best spirit of networking, to ensure proper focus in resource use and synergy in development efforts" (cited in Flor, 2001, p. 16). Flor outlines four major paradigms for analyzing poverty: Technological paradigm, Economic paradigm, Structural paradigm and Cultural paradigm. In the technological paradigm people "believe that the primary cause of poverty is the lack of technological know how in the developing world" (p. 6). Johnson (2005) rightly pointed out that we now live in a world that is infinitely more interconnected than ever before. ICT is the glue that facilitates this interconnectivity, and ICT could be the key to poverty alleviation.

Using Blogs and Robotics in Remote Schools in Outback Queensland

This section draws on research from a large national SiMERR (Centre for Science, ICT, Mathematics education for rural and regional Australia) project for "collaboration between the SiMERR–ICT Australian state/territory hubs and the SiMERR Australian National Centre and aims to raise the awareness of the possibilities for, and impact of social computing on, student learning. This is to be achieved through providing a supported professional learning opportunity for teachers to implement action learning in their own school and participate in a community of practice." (Reading, 2007). A cluster of schools near Longreach in remote western Queensland combined a unit incorporating robotics with professional development via videoconferencing, peer collaboration across four schools using blogs, podcasts and discussion boards. This section will present an overview of the geographical location along with information about the model for implementation of the curriculum unit. It will provide a literature review on robotics and social computing tools and then report on the outcomes from the students' engagement with the classroom unit along with ideas for future directions.

Location and model of implementation

Four western schools situated in a remote Australian outback context in the state of Queensland (Jundah SS, Cameron Downs SS, Prairie SS and Stamford SS) participated in a cross-curricula unit that involved the integration of ICT in the form of social computing tools (blogs, podcasts and videoconferencing) along with robotics. The key teachers were engaged in professional learning and sharing via national videolinks with university academics and other teachers in every Australian state who were also situated in rural, regional or remote locations. In the Queensland model, a key objective was to devise and implement an integrated unit in which the ICT components merged naturally to support and enhance a curriculum unit, rather than a model that introduced the ICT in a stand alone, nonintegrated manner. Planning documents for the unit state that "students will learn the difference between non-renewable and renewable forms of energy, and work in large teams to cooperatively research and produce solutions to improve the energy efficiency of a building in their local community. Students will demonstrate their knowledge and findings to the community, and bring their knowledge to life by programming a Lego robot to complete environmentally conscious activities efficiently, within a short timeframe."

This provided an excellent platform to explore the possibilities afforded by new technologies such as podcasts, blogs, videoconferencing and robotics. Given the disadvantages faced by teachers and students living in isolated areas, the technologies promised remote delivery of professional learning experiences for teachers and the capacity for teachers to share their experiences. It also afforded students the chance to interact and communicate with other students spread across vast spaces in the outback. Teachers participated in a national videoconference where academics in each state presented sessions of a variety of social computing tools. After these sessions, teachers chose from this suite of social computing tools on the basis of "best fit for upcoming units to be completed in the classroom in the current school term." Further videoconferencing sessions involved teachers across Australia discussing their plans for the units and sharing advice and questions about implementation. After these initial sessions, post-unit sessions dealt with sharing outcomes. This presented a massive technical challenge since different states use various types of incompatible or barely compatible equipment. Creativity and innovation on the part of participants, along with use of a videoconferencing bridge at University of New England, led to successful link-ups.

Linking to the literature

Teachers from the cluster decided to explore the use of podcasts and blogs from the social computing tools and also expressed a desire to include the use of robotics. As previously discussed, academics from James Cook University provided professional learning sessions via videoconferencing but also traveled twice to Jundah, west of Longreach, to provide face-to-face professional learning sessions in video and audio capture and editing, blogging and robotics. Teachers were enthusiastic about video editing and podcasting, but ultimately decided to use blogging and robotics in the unit. Therefore this section will concentrate on them.

The Horizon Report (2008) published annually from the New Media Consortium in the United States outlines emerging technologies likely to have a substantial impact on learning, teaching and creative expression. A key 'emerging technology' identified by the group involves simple techniques to publish personal information on the web, especially when that involves collaboration. They point to overarching systems that link this information, and they argue that 'social operating systems will support whole new categories of applications that weave through the implicit connections and clues we leave everywhere as we go about our lives" (p. 4). Although blogs have been described as "a website with dated entries, presented in reverse chronological order and published on the internet" (Duffy and Bruns, 2006, p. 32) or "a weblog is most easily described as a website that is updated frequently with new material posted at the top of the page" (Blood, 2002, p. 12), it must be recognized that blogs are becoming part of a larger system of next generation social networking that has been described as Social Operating Systems (New Horizons Report, 2008).

Lankshear and Knobel (2006, p. 3) rightly point to the ease of use of blogging when they argued that "setting up a blog now simply involved going to a website, signing up for a blog account, following a few fairly straightforward instructions, and in less that 30 minutes one would have some copy up on the web." Ease of use was certainly an attractive aspect that appealed to the teachers in the project. Teachers working on the unit saw blogging, as means of shared reporting and collaboration which would showcase the progress of getting up to speed with robotics. This sharing would occur within the school and across the schools. Lankshear and Knobel recognized this aspect of blogging which they term 'blogging as participatory practice' where "participation means involvement in some kind of shared purpose or activity— taking part in some kind of endeavour in which others are involved" (p. 4).

Witte (2007) described a set of positive outcomes for students engaging in blogging. These were: developing digital fluency, strengthening traditional literacy skills and applying emerging technical skills in a safe and modified environment. Duffy and Bruns (2006) wrote about a different set of complimentary benefits: promoting critical and analytical thinking; promoting creative, intuitive, analogical and associational thinking; increasing access to quality information; and enhancing solitary and social interaction. The teachers agreed that blogging had the potential to contribute to the stated aims of the unit, which coincide with the benefits evident in the research literature.

In addition to the use of blogs, teachers were keen on podcasting as a means of communication and collaboration between the sites and to foster parent and community involvement. Video and audio modes of podcasting were considered as potentially advantageous tools, so professional learning on video capture, editing and file manipulation and conversion to appropriate formats was provided in face-to-face sessions. The Horizon Report (2008) highlights 'grassroots video' as the first emerging technology discussed in the document. They describe the multitude of devices that are now available to record video such as the ubiquitous mobile phones and portable devices along with the facility to quickly and easily upload to sites such as YouTube. According to the authors of the report "once the province of highly trained professionals, video content production has gone grassroots" (p. 10). Although teachers in the cluster recognized the potential of the tool, they decided to purchase more equipment to easily enable audio and video capture and to incorporate this tool in future curriculum units.

A central tool in the integrated unit on nonrenewable and renewable forms of energy was robotics. None of the teachers in the cluster or any of the students had previously had any experience with robotics, so staff from James Cook University provided face-to-face professional learning sessions on Lego robotics at Jundah State School prior to the development of the unit plan. Seymour Papert (1999), a pioneer of educational use of robotics and author of the Logo computer language, has long extolled the benefits of robotics for the development of problem solving skills and a means to introduce high level conceptual development in what he terms a 'real context.' More recently, Johnson (2003) commented on the motivational factors associated with robotics. "There is no doubt that many children and adults find robots fascinating. Sales of affordable robot toys and robot construction sets are reaching unprecedented levels" (Johnson, p. 16). He also concluded that engagement with robotics contributed to the development of teamwork along with conceptual development in the areas of "algebra and trigonometry, design and innovation, electronics and programming, forces and laws of motion and materials and physical processes" (Johnson, p. 18). When he compared robotics to other 'motivating' models of learning he determined that robotics was superior due to its interdisciplinary nature and its capacity to involve more subject areas.

Lau, Erwin and Petrovic's (1999) investigation also examined similar factors but added an element pertinent to this study—the potential of robotics to foster interaction and collaboration between schools. Another study by Matson and De Loach (2004) was the only one that dealt specifically with the advantages of robotics for students in rural and remote areas. Anderson, Timms, and Courtney (2007) found that rural and remote schools were more likely to have teachers teaching ICT without appropriate qualifications and less access to professional development than their city peers. Matson and De Loach agreed and described these schools as being 'underserved.' Their experience across many rural schools in Kansas was that robotics excited the students about science and captured their imaginations. They argued that "students of rural schools must still compete to find a place in an ever increasing global economy dominated by workers with the ability to apply advanced technologies and solve more complex problems than ever before" (p. 2). The experience of high levels of student motivation and excitement about completing a unit with elements of robotics was mirrored in the Queensland outback schools. The notion that robotics leads to increased student interest is easier to support than the contention that robotics increased problem solving ability. Results of research in this area tend to be contradictory with major flaws in the studies such as small sample size, short periods of time for trials, poorly developed instruments with low validity and reliability. One competent mixed method study by Lindh and Holgerson (2005) involving over 700 students over a 12 month period concluded that positive effect sizes were evident for some but not all groups of students.

Professional learning model and results

The professional learning model adopted in the project demonstrates the power of social computing tools to enable collaboration between academics Australia-wide, teachers in rural and remote schools situated in sparsely populated areas spread across the states and also, widespread collaboration with the school clusters. The basic components of the model feature in the diagram below.

Figure 12. Professional learning model employed by the project

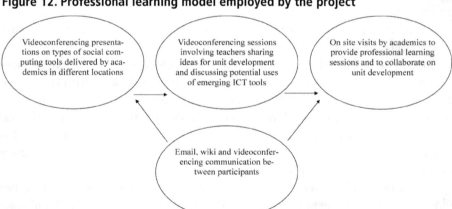

Interviews were conducted with teachers and students to provide some initial insights on how the social computing tools and robotics were received by teachers and students in terms of value.

Positive comments were made about the outcomes in terms of student learning:

In relation to the robotics component the key teacher commented the robotics unit was fantastic. Our students had the opportunity to experience a hands-on real-life project that they had complete control over. It was great to see them working together as a team, supporting and encouraging each other. Since completing the unit, some of my colleagues are now very interested in completing the unit themselves this year.

A typical student comment was

"It was great, we had to work on a really hard challenge but it was fun trying to complete the movements."

This student comment is typical because it indicated that the student had highly positive thoughts about the experience in terms of enjoyment, but also clearly articulated the problem solving challenges and the difficulty.

In relation to responses from teachers concerning the social computing tools, a representative comment was:

"the social computing tools we incorporated in our project enabled my students to expand their horizons within the world of technology. This was fantastic as my students were able to chat to peers from other schools about topics that were relevant to both groups of students. My students' confidence with computer use grew enormously and we noticed the students were more comfortable with using the computers in their free time. The children were emailing each other and visitors and always very eager to check the computers for new emails. As the project progressed, I was excited by the potential of the social computing tools. My school is one of 14 band 5 schools in the Longreach District. All of these schools are geographically isolated and I believe that these social computing tools will be a fantastic way for us to bring our students together around a central curriculum area or unit of study."

"Social Computing will enable me to have my students working in groups with age related peers from other schools. Our students will be able to use these tools to make networks across our district for support in a range of different areas. In many of our schools we may have only 1 student in a particular year level, with the use of social computing tools we will be able to have a number of students from each year level working together on a central topic. This will be fantastic for our students."

When asked about enabling factors for the introduction of social computing tools the responses were:

» The dedication of the teaching and ancillary staff
» The motivation and enthusiasm of the students
» The interest and support of the parents and community
» The interest and support of education district staff
» Being able to tap into larger networks such as other state school systems and academics at different universities

Student responses were favorable, for example:

"We can talk to other students about their ideas with the project we are working on. We can share our ideas and try to make better plans from a shared brainstorm session."

and

"We can look up the answers to questions on the internet or some reference books and share our answers with each other in a forum. We can see what the other kids have said as well."

Teachers identified some challenges of engaging with social computing tools as:

» Lack of knowledge around the use of the tools
» Equipment requirements—what are the latest tools being used?
» Student protection when trying to connect with students from other states

Students and teachers reported tremendous enthusiasm for the robotics component of the project, which culminated in the small and isolated school of Jundah receiving an award in a statewide LEGO robotics competition and producing a movie showing the rapid development of robotics skills to be presented at a community function.

Future Directions

Teachers participating in the social computing project reported on the outcomes to the larger 'Outback Advantage Schools Cluster, and this consequently sparked great interest in expanding the use of social computing tools through shared units across the cluster. An example of an integrating device for the larger cluster from the planning document is: "The Middle School students in Outback Advantage schools will analyse the components of their community (past and present) and make a collaborative decision as the one part that best typifies their place. Once decided, they will design and film a video of this feature. Using Movie Maker, this material will be presented to each community at a Community Event to celebrate the advantages of Outback communities. The Early Years students will look back over their life so far and talk about experiences that are important to them. They will design and produce a pictorial presentation that documents these experiences with captions and/or commentary to present at the Community Event." Another unit heavily integrates podcasting as a way of enhancing the unit. Social computing tools have been taken up very enthusiastically by the cluster of schools and this seems to be a strong indication that these tools are particularly useful to schools facing the challenge of geo-graphical isolation. More research needs to be conducted to determine the benefits of these new technologies, which might assist in ensuring that rural and remote students have the same opportunities for high quality and motivating educational opportunities as their urban counterparts.

...

PART 5

Software and Hardware Developments

The Powerful Coupling of Open Source Software and Low Cost Hardware

Open Source Software (OSS) offers free, open and modifiable code to developers and has been flagged as a potential solution to the problem of software cost restrictions on users in developing countries and for low earning groups in developed countries. It has also been taken up enthusiastically by users who can afford commercial software but recognize the benefits of OSS. One well known example is the growing use of Mozilla Firefox web browser and the success of its predecessor Netscape. This burgeoning use of Open Source Software has profound implications for equitable use of ICT in education. For example, the Indonesian government in 2004 launched the IGOS project—Indonesia Goes Open Source. Under this project OSS was to be used in all government offices, followed by a roll-out of OSS in other institutions, including all government schools (Rennie, 2004). Widespread use of OSS in government and business sectors would be likely to influence software use in the home, thereby extending the OSS effect on education in formal, informal and business environments.

A powerful combination has emerged where new low cost laptops are being produced and packaged with the OSS operating system Linux and a suite of easy to use software such as Open Office. It is this relatively new combination of free or low cost software and low-cost hardware which also provides web and networking connectivity that is truly transformative. One commercial example of this combination is the ASUS Eee mini-laptop and a large philanthropic example entitled the One Laptop Per Child project initiated by Nicholas Negroponte and launched at the Tunis United Nations World Summit on the Information Society in November 2005. This section will examine definitions of OSS and look at the question of motivation for the open source movement, particularly in relation to programmers and will then outline the benefits and challenges of OSS and conclude with some details about the existing and emerging couplings of low-cost hardware and OSS and the One Laptop Per Child project.

Defining Open Source Software

Haruvy, Prasad and Sethi (2003, p. 381) define OSS as "the use of and contribution to a shared source code that can then be freely redistributed and reused in components of code. The open-source paradigm lends itself to collaborative software development by a world-wide community of developers to build software, indentify and correct bugs, and offer enhancements." Bonaccorsi and Rossi (2003, p. 1243) argue that "open source software can be analysed as a process innovation: a new and revolutionary process of producing software based on unconstrained access to the source code as opposed to the traditional closed and property-based approach of the commercial world." The official site of OSS has a lengthy definition that sets boundaries for projects under the open source banner:

> Open source doesn't just mean access to the source code. The distribution terms of open source software must comply with the following criteria:
>
> ### 1. Free redistribution
> The license shall not restrict any party from selling or giving away the software as a component of an aggregate software distribution

containing programs from several different sources. The license shall not require a royalty or other fee for such sale.

2. Source code

The program must include source code and must allow distribution in source code as well as compiled form. Where some form of a product is not distributed with source code, there must be a well publicized means of obtaining the source code for no more than a reasonable reproduction cost, preferably downloading via the Internet without charge. The source code must be the preferred form in which a programmer would modify the program. Deliberately obfuscated source code is not allowed. Intermediate forms such as the output of a preprocessor or translator are not allowed.

3. Derived works

The license must allow modifications and derived works and must allow them to be distributed under the same terms as the license of the original software.

4. Integrity of the author's source code

The license may restrict source code from being distributed in modified form *only* if the license allows the distribution of "patch files" with the source code for the purpose of modifying the program at build time. The license must explicitly permit distribution of software built from modified source code. The license may require derived works to carry a different name or version number from the original software.

5. No discrimination against persons or groups

The license must not discriminate against any person or group of persons.

6. No discrimination against fields of endeavor

The license must not restrict anyone from making use of the program in a specific field of endeavor. For example, it may not restrict the program from being used in a business, or from being used for genetic research.

7. Distribution of license

The rights attached to the program must apply to all to whom the program is redistributed without the need for execution of an additional license by those parties.

8. License must not be specific to a product

The rights attached to the program must not depend on the program being part of a particular software distribution. If the program is extracted from that distribution and used or distributed within the terms of the program's license, all parties to whom the program is redistributed should have the same rights as those that are granted in conjunction with the original software distribution.

9. License must not restrict other software

The license must not place restrictions on other software that is distributed along with the licensed software. For example, the license must not insist that all other programs distributed on the same medium must be open source software.

10. License must be technology-neutral

"No provision of the license may be predicated on any individual technology or style of interface" (Coar, 2006, n.p.).

These criteria have been designed to protect the ethos of open source development and to provide an explicit framework for OSS projects to work within. Some OSS developments have been spectacularly successful. Ven and Mannaert (2008) have identified that the most mature and successful OSS projects mainly consist of platform software such as Linux, Bind, Apache and Sendmail. Other examples of successful OSS include PHP and Pearl programming languages, Mozilla Firefox and MySQL database. Von Krogh and Spaeth (2007, p. 237) point out that "over the last 15 years, many open source products have made successful inroads into these segments attracting many millions of users. For example, in 2005, Apache achieved a 60% market share for web server software, In the same year, Firefox, the browser, achieved a 13% market share and turned over more than 50 Million USD for the Mozilla foundation that markets and coordinates its development."

Open Source Software Development and What Motivates Developers?

Steps used by OSS developers to complete projects vary one project to the next, but a common misconception is that OSS development is ad hoc and undisciplined. Many projects, especially larger ones, follow a common set of development procedures and exhibit disciplined, yet open and flexible approaches. Barcellini, Detienne and Burkhardt (2008, p. 559) describe "people connected together on the Internet with a common goal—to develop software—with the 'meta'—objective of producing and constructing knowledge about the artefact they develop for the benefit of the community. Their activities are framed by implicit and explicit rules: volunteer participation or evaluation of work by a peer review mechanism, for instance (e.g. Raymond, 199). Major OSS projects are highly hierarchical and meritocratic communities."

Bonaccorsi and Rossi (2003) outlined steps often followed when projects emerge:

a)　An individual or organisation have a problem or need for a particular type of software
b)　The individual or group may seek others with the same need or concern
c)　Form a group that starts working on the project
d)　After working on the project for a while, the group seeks wider feedback from the online community through Internet communication such as e-mail lists, newsgroups, blogs, etc.
e)　The project attracts further participation

Like Barcellini et al., they debunk the notion of 'anything goes' in OSS development and point out that "most successful Open Source projects, far from being anarchical communities, display a clear hierarchical organisation. On the other hand, contrary to what happens in the proprietary paradigm, the roles within the hierarchy are not strictly assigned" (Bonaccorsi & Rossi, 2003, p. 1247). Much of the commentary on OSS development centres on the importance of the emergence of an effective 'creative leader' in successful projects. If the process is sometimes quite formal and arduous, why then do programmers engage in this process, often in their own time and without any immediate financial benefit?

Economists have been puzzled at the success of OSS development, as they doubted that altruism alone would provide enough motivation to sufficiently engage serious programmers for lengthy periods of time. Although altruism can be a strong influencing factor, studies across different disciplines have uncovered a variety of motivating factors (Bonaccorsi & Rossi, 2003, Krogh & Spaeth, 2007). Bonaccorsi et al. argue that "altruism might at most explain the behaviour of people writing software in their spare time but not the behaviour of those who have devoted considerable resources of time and intellect, (p. 1245). A summary of the motivating influences cited in the studies is:

a) Developing new software provides intrinsic motivation via satisfying the intellect in a similar manner to scientists making new scientific discoveries

b) Creating the software fills a commercial or personal need not catered for by existing software

c) Free programs on the web can be linked to advertising, thereby creating future revenue

d) Sharing results can lead to improvements via feedback from other members of the OSS community

e) Programmers gain recognition and prestige through their successful OSS efforts, thereby increasing their human capital worth, e.g., they may become more attractive to ICT firms

f) Many programmers regard their work as an art form and gain pleasure through a creative outlet

g) Programmers can learn new skills by openly collaborating with others in a global network

h) OSS programmers develop a sense of kinship or affinity groups and contribute in a similar way to family members contributing to the health of the family

i) The OSS ethos attracts the attention of ICT researchers since its processes, products and underlying code are not kept secret in the way that commercial software development is

j) Sharing modular development with the OSS community is sensible since upgrades to OSS operating systems such as Linux can then take the new modules into account when upgrading to new versions

Negroponte and the One Laptop Per Child Program

Nicholas Negroponte and the UN Secretary-General in 2005, Kofi Annan, launched the prototype of the US$100 laptop at the World Summit on the Information Society (WSIS) in Tunis. The computer labelled the 'Green Machine' was a prototype for an ambitious project under the title of One Laptop Per Child (OLPC). Negroponte explained the rationale of lowering the cost of hardware by manufacturing a low cost, relatively small display that would still have adequate performance in conjunction with cutting the excess from bloated commercial software and hardware products and replacing the hard disk with solid state chips. Another cost reducing feature outlined was the voluntary design and project management by many of Negroponte's colleagues, particularly the group at MIT and recruits from the ICT industry. Reducing costs by large scale manufacture was also a key strategy outlined by Negroponte. The ability to power the laptop via a small hand crank was also a feature of the original prototype along with built-in networking (including wireless and mesh) capabilities. Since the machine was designed to fulfil a vision of providing a window to a world for children in developing countries, it was essential to allow use of the machine where mains power was intermittent, not always available to all or nonexistent.

More recently, Battro (2007) outlined on the OLPC website that the original prototype devised by Design Continuum has now been modified by a new team at Fuseproject. They have achieved the specification of a low cost (US$35) laptop screen that has the ability to switch to a high resolution black and white display for reading e-books in high light situations. The rugged machine has long life batteries and connectivity via multiple USB ports and is equipped with flash memory (no moving parts such as a hard disk), wireless and mesh networking built in and innovative power source solutions. Battro cast doubt on the original hand crank as being less than efficient and flagged the development of separate power generating equipment. The efficiency required for power generation was approximately one minute cranking for 10 minutes of computer use and the original hand crank failed to provide the necessary efficiency levels. Another feature of the laptop is its resistance to damage by dust, sand and water. Although desktop computers have the potential to be manufactured more economically than laptops, mobility was seen as a critical component of the OLPC project. Tabb (2008) argued that "the designers of the laptop itself seem to have thought of almost every possible feature necessary for its use, including durability, child-friendliness, toxicity and outdoor usage."

Battro (2007, n.p.) points to the UN Millennium goals of 2000 as inspiration for the project, particularly:

> "19. To ensure by the end of 2015, children everywhere, boys and girls alike, will be able to complete a full course of primary schooling and that girls and boys with have equal access to all levels of education.
>
> 20. To ensure that the benefits of new technologies especially information and communication technologies…are available to all."

He describes the widespread provision of laptops as a 'digital vaccination' implying that the benefits will 'cure' or improve poverty and lack of education through access to ICT. While it is essential that children have access to ICT, it should also be remembered that access alone does not automatically lead to benefits. This must be provided along with programs that support and encourage young people to use computers in productive ways. Negroponte acknowledges the influence of mathematician and educationalist Seymour Papert's ideas on the development of the project. Papert has been a staunch supporter of using computers in particular ways that extend the learners' ability to solve problems, to create and to communicate. While communication ability has been placed at the forefront of ICT's benefits, Papert sees the most important advantage as extending young people's ability to engage in problem solving and the creation of products. The success of OLPC depends equally on provision of resources and the provision of powerful ideas and support for the productive use of the equipment. Tabb (2008) points out that Negroponte labels OLPC as an education project and not a laptop project but that the Indian Education Secretary, Sudeep Banerjee, was not convinced and described the project as being pedagogically suspect and called on funds to be used for more teachers and classrooms.

Battro (n.p.) argues that "most experiences of giving computers to schools or furnishing computer labs in educational institutions have not been able to bridge the digital gap. The only way to achieve this goal is for the child to own the computer, to own a light robust portable computer, a laptop with the lowest possible energy requirements, capable of bypassing standard electrical restraints using mechanical or solar energy devices, that he or she can carry home." Although the provision of the portable computer offers potential for bridging educational divides, it certainly cannot guarantee it, as shown by a recent study by Malamud and Pop-Eleches (2008). In this study the researchers compared families in Romania selected by the government for the provision of computer vouchers. Since about half the eligible families missed out on the voucher, the researchers were able to compare educational results of the two similar groups. The children receiving the computers in 2005 were shown to have lower educational attainment and aspirations in 2007 and significantly, had little or no educational software installed on the computers and, reported low or no use of the computers for educational purposes. The study concluded by highlighting the importance of direction from parents or carers in supervising and modelling effective use of the ICT resources.

These results are in accord with Wenglinski's (1998) U.S. national study which provided evidence that particular types of computer use can lead to higher educational outcomes and other uses can have the opposite effect. Zeal for provision of one laptop per child should be tempered by the realisation that this is the first essential step in the process. Battro does make valid points about the lack of books in developing countries and the potential of digital books delivered via the devices, the potential for health education and that children equipped with computers can be teachers as well as learners—acting as peer tutors for adults and other children.

Visions associated with programs such as OLPC are, by necessity, usually idealistic. For example, the recent keynote speech from the OLPC countries meeting at MIT in May 2008 spoke of the 'family passion' "of Nicholas Negroponte and his team to give 'one laptop per child' to children in need...Our purpose is simple. Our purpose is to give every single child in the developing world and the developed world, the power to escape poverty, the power to escape ignorance, the power to escape years of neglect, deprivation and non-fulfilment. By doing so we give them a life chance, an opportunity, a window to realise their life potential and join the whole of the global community whatever their circumstances." The keynote goes on to present a vignette "a hut in an African village. An old man and an old woman cooking their dinner by firewood. At a small table a young boy about 11 years of age is tapping into a computer looking at the keyboard in the half light of the flickering kerosene lamp. He has escaped his hut. His mind is elsewhere connected to the global internet community. He is doing his homework together with 7 other kids in similar huts 1 km away. The children are in nodes and the village community is brought together by their laptops."

Positive hype to bolster support for OLPC has met with criticism such as the following text from a blogger (http://detailsaresketchy.wordpress.com/2007/05/21/cambodia-one-laptop-per-child/)

> "Lesley Stahl checks in on the current state of Nicholas Negroponte's 'One laptop per child' project. Actually, it sounds a lot less like reporting, and a lot more like she has just reprinted Negroponte's bombast. 'The first English word of every child in that village was Google,' [Negroponte] says. The village has no electricity, no telephone, no television. And the children take laptops home that are connected broadband to the Internet."

When they take the laptops home, the kids often teach the whole family how to use it. Negroponte says the families loved the computers because, in a village with no electricity, it was the brightest light source in the house.

> "Talk about a metaphor and a reality simultaneously," he says. "It just illuminated that household."

Once the computers were there, school attendance went way up.

Negroponte says that in Cambodia this year 50 percent more children showed up for the first grade because the kids who were in first grade last year told the other kids, school is pretty cool.

> **"Attendance went up nationwide because one village got 50 laptops? Please."**

Other bloggers provide supporting arguments concerning pilot schemes in various countries:

> "I work with Cambodians from villages all over the country who have gone to school and now speak English better than some immigrants who have been in the U.S. for 30 years. And they are earning more than a $1,000 per month. I work with a girl who learned English from a monk in her village. I've told a 7 year old the ABC's three times and she recited it to me 3 weeks later when I saw her again."

If Cambodians have access to information and education they can begin to try, even if it doesn't get them too far, to make a better life for themselves and their families. There may be causes that could be more beneficial to the people. While giving laptops is like sticking your finger in a leaking dam it could be something for some.

You know books are not even hardly available in Cambodia. Most of that was destroyed. So there is not too much university oriented educational information to be had from within the country. The easiest way is for them to learn English.

So yes you are probably correct to say that attendance has probably not gone up in the whole country. I guarantee you if I took my own laptop, just one, down to a Cambodian village, there would not be a kid who would want to miss it! And it fosters their interest in learning. That may not seem like much to you, but in a country like Cambodia it is essential to their hope for prosperity however small it might be.

And while you may pass through a village and think they are all stupid and should be taught to ride a motorbike better, there are people there who didn't have 50 cents to buy a pencil ten years ago. And there wasn't a bookstore down the street. They were isolated from the world for almost 25 years.

So while these adults are illiterate it's not because they chose to be. They would have been killed if they were educated not too long ago.

The first step for Cambodians to begin to help themselves is through education."

Another blog site documented the response of local teachers to OCPL in Arahuay in Peru (http://radian.org/notebook/astounded-in-arahuay). Three major changes had been noticed by the local principal and teachers:

"Mr. Navarro and Mrs. Cornejo spoke amongst themselves for a few minutes. Then Mr. Navarro said they agreed there were three key changes.

As there are few roads in and around Arahuay, the children don't communicate much outside of school—with anyone. The teachers started independently pointing out to Mr. Navarro that this was changing once the laptops arrived: kids started talking to each other outside of school hours over the mesh, and working together more while *in* school. They started talking a lot more with each other in person, and conquered their previously paralyzing fear of strangers.

The second thing, Mrs. Cornejo jumped in, is that the kids used to be pretty selfish, an unsurprising consequence of the abject poverty in much of Peru. It's not that the kids are starving, it's just that they don't *have* very much; what they do have, they're reluctant to share. With the laptops, the kids had to turn to each other to learn how to use them. Then they realized it was easy to send each other pictures and things they've written—and it became commonplace. The sharing, asserts Mrs. Cornejo, extended into the physical world, where once jealously-guarded personal items increasingly started being passed around between the kids, if somewhat nervously.

'Finally,' opened Mr. Navarro, and hesitated. He gave me another long look, clearly unsure if to proceed. I put on my best smile, and assured him it's exactly the things he would hesitate to tell me that I want to hear most. He cleared his throat, and in a conspiratorial, low voice—despite the fact we were in an empty room in the town hall—explained he was sure, in the beginning, the pilot would fail.

'Children's fathers used to seethe with fury when the laptops were passed out, because the kids no longer wanted to help work in the field all day,' he continued.

Mr. Navarro speaks in slow, measured sentences. He is thoughtful and confident, both reminders—along with his weathered face—of being, for many years, foremost a teacher.

'I didn't know how we'd stop the fathers from revolting and making the kids return their XOs,' he says, shaking his head slightly. 'The kids solved the dilemma for me: they taught their fathers how to use the Internet and a search engine.'

'Then they started showing them the work they were doing for school. The reports they wrote, the pictures they took, the notes they compiled. And the fathers had actual proof that their kids were learning,' he concluded."

The principal had some initial concern about the effect of the laptops on students' behaviour. According to the report "both Mrs. Cornejo and Mr. Navarro thought the XO would exacerbate some existing discipline problems at the school. One student, whose name I'll withhold, commonly gets in fights with others, didn't speak to or play with his classmates, and would normally sit in a corner of the classroom by himself. The principals anticipated the XO would make him even more territorial and isolated, but they were taken by complete surprise when he became the first kid to figure out the laptop, and then started teaching the others, who curiously flocked around him.

'We don't tell these feel-good stories, these fairy tales,' Mr. Navarro responded to my unspoken skepticism. 'It's just what happened. It's just how it is.'

I headed back to Lima several hours later, astounded by what I heard and saw."

Indeed the core of the success of the project to date centres on South America. Tabb (2008) outlined the current orders for green laptops as: "Peru, 40 000 with an option to buy 210 000 more. Brazil, 150 000, Uruguay, 100 000, and Carlos Slim of Mexico has individually ordered 50 000 laptops. It is also not surprising that once one country in a region makes a purchase, others may be more inclined to follow suit."

Felsenstein (2005) who was involved with the Jhai Project (also described in this chapter) critiques OLPC as being top driven. He claims that "by marketing the idea to governments and large corporations, the OLPC project adopts a top-down structure. So far as can be seen, no studies are being done among the target user populations to verify the concepts of the hardware, software and cultural constructs. Despite the fact that neither the children, their schools nor their parents will have anything to say in the creation of the design, large orders of multi-million units are planned" (http://fonly.typepad.com/fonlyblog/2005/11/problems_with_t.html).

Perhaps the top-down model and other challenges contributed to the lower than expected uptake. In an effort to gain momentum for the project, a decision was made to make machines available to buyers in the U.S. if the purchases could leverage the supply of the laptops to children in developing countries. A report on BoingBoing stated that "the Foundation that manages the One Laptop Per Child Program (which will give one low-cost, Linux-based laptop to every child in the developing world— eventually) is making their machines available for sale in the developed world this Christmas. The price is $399, and includes two laptops, one of which will be given to a child in the developing world. I've just signed up to get one—I just wish that this was structured as a donation to the Foundation, since I think they'd sell a ton of these if the purchasers could get a tax receipt for them just before the tax year closes" (http://www.boingboing.net/2007/09/23/one-laptop-per-child-1.html). Tabb (2008) reported that later offerings did follow the donation structure which allowed the tax deduction.

No doubt, positive reports from trial programs would be helpful in recruiting industry partners and securing the services of volunteers. This would be essential for the success of a large project fuelled mainly by goodwill, where many people provide in-kind or financial resources. It also needs to operate in an environment where commercial ICT interests might regard the project as a threat. Indeed, Negroponte expressed concern about the actions of chip manufacturer Intel after OLPC designed the laptop utilising an AMD processor. In an interview aired on the television program *60 Minutes* Negroponte described Intel's actions as predatory when they provided alternative low cost laptops to Mexican students in opposition to the OCPL program (http://www.olpctalks.com/nicholas_negroponte/olpc_60_minutes_interview. html).

Open Source Software for OLPC

Benjamin Hill (2007), a member of the OLPC team, outlined the open principles adopted by the project and argued that the project could only create a means to freedom and empowerment if it embraced openness. These are the principles posted on the OLPC wiki:

» "Must include source code and allow modification so that our developers, the governments that are our customers, and the children who use the laptop can look under the hood to change the software to fit an inconceivable and inconceivably diverse set of needs. Our software must also provide a self-hosting development platform.

» Must allow distribution of modified copies of software under the same license so that the freedoms that our developers depend upon for success remain available to the users and developers who define the next generation of the software. Our users and customers must be able to localize software into their language, fix the software to remove bugs, and repurpose the software to fit their needs.

» Must allow redistribution without permission—either alone or as part of an aggregate distribution—because we can not know and should not control how the tools we create will be re-purposed in the future. Our children outgrow our platform, and our software should be able to grow with them.

» Must not require royalty payments or any other fee for redistribution or modification for obvious reasons of economy and pragmatism in the context of our project.

» Must not discriminate against persons, groups or against fields of endeavor. Our software's power will come through its ability to grow and change with the children and in a variety of contexts.

» Must not place restrictions on other software that may be distributed along side it. Software licenses must not bar either proprietary, or 'copyleft' software from being distributed on the platform. A world of great software will be used to make this project succeed—both open and closed. We need to be able to choose from all of it.

» Must allow these rights to be passed on along with the software. This means that we must not provide a license specific to the $100 Laptop project or organization or its customers. While we are the developers of this platform today, the users of this platform are the developers of tomorrow and it is through them that the platform will succeed, be transformed, and be passed on. They need the same rights as we do.

> » Must not be otherwise encumbered by software patents which restrict modification or use in the ways described above. All patents practiced by software should be sublicenseable and allow our users to make use or sell derivative versions that practice the patent in question.
> » Must support and promote open and patent unencumbered data interchange and file formats.
> » Must be able to be built using unencumbered tools (e.g., compilers)."

Hill emphasises that these principles are necessary given the current scope of global diversity and to allow for future developments that will inevitably supersede the current project. Although OSS has greatly assisted the project by reducing costs, it cannot overcome the cost of broadband access in many developing countries, particularly Africa, where connectivity costs are especially high.

Commercial Low Cost Hardware and Open Source Software Combinations

Negroponte, during a *60 minutes* interview, outlined his opinion that commercial producers of rival chip manufacturers viewed his large scale project as a threat and responded accordingly by offering alternative products at below cost. This was perceived as a measure to stop OLPC getting a foothold in certain countries. It is reasonable to assume that any mass production of lost cost laptops would be seen as a threat to other laptop manufacturers. Likewise the promotion of Open Source Software to a mass market would be seen as threat to software development companies such as Microsoft.

An interesting recent development that may have a greater impact than OLPC on increasing the number of portable computers in poorer communities has been the realisation that the principles of OLPC could also be applied to commercial production of laptops. Ideas flagged by the OLPC project such as smaller, cheap screens; replacing hard drives with solid state memory chips; minimum specifications apart from wireless connectivity; and OSS software have been embraced by some leading PC manufacturers. In the case of ASUS, it has led to great commercial success, since the light, portable machines can be sold at a very low cost in comparison to previous mini-laptops.

Taub (2008) described mini-laptops, each with a similar makeup of smallish screens and low retail prices, such as the ASUS Eee PC900; Everex Cloudbook; Hewlett-Packard 2133 mini-note; and the Fujitsu Lifebook U810. Two of these models, the ASUS and the HP, offer Linux as the operating system and Open Office software as standard. Interestingly, the previous model by ASUS only offered Linux but the new version with a slightly bigger screen can be purchased with MS Windows as the operating system. It's a reasonable assumption that Microsoft offered ASUS an excellent deal to ensure the mass market with Linux was reduced. The success of the ASUS model is evident by the hundreds of thousands of the Eee that have been shipping since its release, and the end result is a reduction in the high cost barrier associated with laptops and a step in the right direction to erode the digital divide, provided that low cost hardware and software can be coupled with support to scaffold particular ways that computers can be used to enhance learning. In another interesting development the latest ASUS model offers a choice between Linux or Windows XP, indicating that Microsoft offered ASUS an attractive deal on the Windows operating system.

Computer-Based Games and Equity— The Rise of 'Serious' Games

In recent times there been a growing recognition of the potential of computer-based video games to enhance learning. Gee's (2003) seminal work, "What Video Games Have to Teach Us About Learning and Literacy" outlined 36 learning principles that he distilled from the elements of game play of popular video games designed for entertainment. There has been a surge in video game design and production for the purpose of teaching specific skills, especially for people with particular needs. The 2007 European Conference on Game-Based Learning showcased some impressive new developments in these types of games that are now often referred to as 'serious games.' Perhaps the most impressive examples using the most sophisticated game engines and artistry were unfortunately developed for the U.S. military, such as the photo realistic simulation of a nuclear submarine that enabled crew members to totally familiarize themselves with the unfamiliar environment before boarding the vessel and the role playing scenarios of different environments in Iraq. These featured photo-realism with every detail of blades of grass or grains of sand and varied from combat scenarios to approaching residents of houses in culturally sensitive ways, using appropriate language.

Serious Games for Students with Disabilities

Although not as quite as sophisticated, but just as impressive, was the work presented by David Brown from the Interactive Systems Research Group (ISRG) at Nottingham Trent University concerning virtual environments for disengaged youth with severe learning disabilities. Brown, Battersby and Shopland (2005) had previously written about their development of role playing games for people with intellectual disabilities to practice valuable skills within a safe and motivating context. Their more recent game presented at the conference was aimed at promoting the social inclusion of excluded learners while it provided lessons in development of positive self-esteem, managing aggression and responding appropriately to stress (Brown, Shopland, Battersby, Lewis & Evett, 2007). The target audience included those:

» Permanently excluded from school
» Long term ill
» Drug and alcohol dependent
» Young offenders
» Referred by social services
» Having severe behavioural difficulties
» Having severe literacy and numeracy difficulties
» Being in care
» At risk of harm
» Being refugees

The trial of the game Quest produced encouraging results with the researchers concluding that "these results are encouraging enough for us to suggest an augmented experimental design to measure engagement based on video methods and automatic coding and analysis of behaviours" (Brown et al., 2007, p. 45). It was expected that some of the participants with severe behavioural problems would not engage with Quest for long, but the program maintained the interest of all participants.

Brown and colleagues' prior work with the creation of virtual environments for students with intellectual disabilities included an interesting scenario where participants learn how to negotiate road crossings and use public transport (Brown, Battersby & Shopland, 2005). Standen and Brown (2005, p. 272) summarized and reviewed the advantages of this technique in developing social skills in a safe environment.

> Virtual reality (VR) possesses many qualities that give it rehabilitative potential for people with intellectual disabilities, both as an intervention and an assessment. It can provide a safe setting in which to practice skills that might carry too many risks in the real world. Unlike human tutors, computers are infinitely patient and consistent. Virtual worlds can be manipulated in ways the real world cannot be and can convey concepts without the use of language or other symbol systems. Published applications for this client group have all been as rehabilitative interventions. These are described in three groups: promoting skills for independent living, enhancing cognitive performance, and improving social skills. Five groups of studies are reviewed that utilize virtual technology to promote skills for independent living: grocery shopping, preparing food, orientation, road safety, and manufacturing skills. Fears that skills or habits learnt in a virtual setting would not transfer to the real world setting have not been supported by the available evidence, apart from those studies with people with autistic spectrum disorders. Future directions are in the development of more applications for independent living skills, exploring interventions for promoting motor and cognitive skills, and the developments of ecologically valid forms of assessment.

After reviewing the literature on the use of virtual reality (VR) in the areas stated above, the researchers concluded that "VR should not be seen as a solution that would suit all users, and currently there are limited applications that are designed for those with visual impairments. However, the studies reviewed all indicated the acceptability of this medium for many users, and potential applications are continually proposed" (p. 281). The review confirmed that people with intellectual disabilities were amongst the most affected by social exclusion, thereby one of society's most vulnerable groups. It was also found that students with intellectual disabilities were often denied real-world experiences and did not have the same opportunities to practice skills which would lead to independence. VR and role playing games offer people with disabilities the chance to practice activities such as learning how to use public transport, how to do the shopping, how to prepare food and many other useful skills, in an environment where participants feel safe and willing to take risks.

Healy and Connolly (2007, p. 105) defined games-based learning (GBL) as "the use of a computer games-based approach to deliver, support and enhance teaching, learning, assessment and evaluation." Healy and Connolly (2007) argue that there is a great need for further research to evaluate the impact of GBL on learning, emphasising a particular need for longitudinal studies. They contend that there needs to be a greater social acceptance of games as an effective learning medium in schools. They cite research by TEEM (2002) that supported the notion that parents perceive GBL as contributing to learning in home contexts. In this study 85% of parents believed that GBL had contributed positively to their children's learning. Healy and Connolly concluded from their study that the learning experience of many students could be enhanced through the use of GBL and advocated a hybrid model involving traditional methods in combination with GBL.

Another potential use of GBL involves virtual environments as a context to learn about indigenous cultures. The rest of this section will concentrate on this aspect.

Gee (2007) argues that video games can lead to deep intellectual engagement in a highly motivating context. Often the context for engagement with video games has been described as occurring within a 'popular culture.' This section examines the benefits of a proposal for video game development in Australia to introduce learners to aspects of specific, traditional Aboriginal and Torres Strait Islander culture. This melding of popular and traditional culture mirrors the work of the University of Wisconsin, Games and Professional Practice Simulation group in relation to American First Nation culture. To contextualize this for readers, some details of indigenous Australian cultures and the sites of the study are described. The literature review looks at what is known about the advantages and disadvantages of educational games and links this to the emerging use of games to foster knowledge of indigenous cultures.

GBL and Indigenous Culture

Description of the Project

In 2007, a group of academics from James Cook University, Australia, applied for seed funding to explore the feasibility of this idea. The project was described in the application as:

"The Computer Game Exploration: Computer gaming is an emerging 'new art form' (Gee, 2005, p. 61), a "visual-motoric-auditory-decision-making symphony, and a unique real-virtual story" in which player and game designer 'co-produce' a new form of 'performance art.' Computer games allow players to 'feel like heroes' in their own lives and communities, promoting connectedness between their own stories and the stories of 'others' through (massively) multi-player on-line role-player games (MMORPG). Games are also semiotic spaces; they are indelibly connected to 'scientific' explorations of problem solving—control, efficacy, agency and meaningfulness (Gee, 2007). Over the last few years a new field has emerged around the idea of 'playful learning,' suggesting that the unforced learning opportunities arising from 'play-based learning' can engage learners. Computer games are the agents of new forms of literacy; here play can mean reading, writing and/or programming in an environment where text consumption and production are immediately and consequentially connected. Play connects text consumption and production, aligning language and learning in a way that 'fits' with how the human brain is designed to learn (Prensky, 2001). Playful learning further distributes intelligence via the creation of 'smart tools' such as games and robots, building learner capacity through collaborative 'cross functional affiliation' through team and multiplayer collaborations (Gee, 2007). Computer gaming can empower learners to assume 'control and mastery' in their own stories, play and lives. This emerging sense of control can further enhance scholastic, individual and community well-being.

This project brings together a panel of experts in the form of an extended and open discussion 'hub' to explore how 'interactivity' and the 'customization' of play environments can contribute to a form of playful learning that delivers 'strong identities' and 'well-ordered problems' (Gee, 2007). The task here is to unpack how games might work in the lives of indigenous learners to mediate tensions between learning that is 'doable' and learning that is 'frustrating'; between problems that are 'well ordered' and problems that are 'decentring'; between learning that validates an existing cycle of community expertise, and one that undermines it; and between a form of 'mentoring' that enables full immersion without compromising guidance, instruction and identity. The project builds on the corollary between 'play' and 'science'— both require models to facilitate thinking, learning and knowledge-building in innovative and creative ways where the design of one can deeply transform the other. The panel will explore the use of computer games to inform the development of a project to research interventions to move students beyond 'learning about' the world to how these technologies can support 'learning to be' in it. Digital video will be the medium through which panel interactions are captured and bracketed for archive and distribution.

The project will build capacity by establishing a cross-disciplinary project reference group in the form of an interdisciplinary discussion hub (panel of experts) comprising experienced and early career researchers from James Cook University (Schools of Education & Indigenous Australian Studies), international contributors—staff from the Edmund Rice Flexible Learning Centre Townsville, Far North Queensland Indigenous Education Workers, and local community members" (JCU ARACY Submission, 2007).

Aboriginal and Torres Strait Islander culture

A major challenge of representing Aboriginal or Torres Strait Islander culture in a video game is the tremendous diversity that exists in these cultures, depending on the particular location in Australia. Over 200 different indigenous languages have been documented, along with details of unique aspects of different communities. Although (according to Australian Census statistics) the majority of indigenous Australians now practice some form of Christianity, many Aboriginal and Torres Strait Islander people have a strong traditional link to the land. Aboriginal people often identify with the 'Dreamtime,' which explained the ancient time of creation and the setting of traditions and ways of being for contemporary times. Torres Strait Islander people have a Melanesian cultural background and often identify strongly with the ocean, the Torres Strait Islands (between the Australian mainland and New Guinea), seafaring ways and agricultural traditions. It is not possible to provide more than a cursory overview of these rich and interesting cultures in a brief section, but it is important to point out that any attempt to represent these cultures in video games would need to be specific to a particular geographic area and associated culture and would need to include significant consultation with indigenous people, particularly elders to ensure authenticity and genuine collaboration.

A starting point for investigation to further the ideas outlined in the project will focus on the Aboriginal people of Groote Eylandt and the Torres Strait Islander community at Moa Island. Groote Eylandt is an island 630 km from Darwin, in the northern top-end of Australia. The population is just over 2000 people, many of whom belong to the Anindilyakwa tribal group. It is the largest island (2687 square kilometers) in the Gulf of Carpentaria. Moa Island is 90 km from the main settlement in the Torres Strait, Thursday Island. It is divided into two communities, St Paul's and Kubin. The community at St Paul's consists of 210 people in an area of 1770 hectares. These two communities have a history of cooperation with the university and the researcher. This past history of collaboration and the geographical isolation of both communities offer advantages for a pilot feasibility study.

Advantages of the games approach

This section will outline some of the possible advantages for students, teachers and the community that may be afforded by a games-based approach to introducing traditional Australian indigenous culture. Of primary concern to many educators is finding pathways to improved student learning in general. Van Eck (2006, p. 18) pointed out, "studies that use rigorous statistical procedures to analyse findings from multiple studies (meta-analysis), have consistently found that games promote learning and/or reduce instructional time across multiple disciplines and ages." Van Eck cites Szczurek (1982) and Van Sickle (1986) and Randel, Morris, Wetzel and Whitehill (1992) as examples of meta-analysis. Certain factors relevant to enhanced student learning outcomes to be discussed further are motivation, immersion, informal learning outside of the confines of institutions, deeper learning and learning about other cultures in context.

Guynup and Demmers (2005, n.p.) emphasise the value of play in relation to positive motivation. They contend that "given the motivational issues surrounding the challenge of learning, it is clear that some types of elements must be added to create a sense of play." An element of play is seen to be crucial to include as a design emphasis when creating a video game featuring indigenous culture. Van Eck (2006), like Malone and Lepper (1987), believes that the inclusion of a fantasy element makes games intrinsically motivating. For the target audience of the game, taking part in a cultural/adventure experience on a remote tropical island would provide a substantial fantasy element. Immersion is associated with high levels of interest and motivation. Van Eck refers to the principle of 'flow' posited by Csikszentmihalyi when he explains that flow results through concentrated engagement in an activity when a high level of immersion causes the player to lose track of time and the goings on of the outside world. This often results in performance at an optimal level. Kardan (2006), a pioneer in developing video games with embedded traditional cultural knowledge, argues that computer role playing games can provide a powerful medium for the teaching of particular subjects in a way which can be engaging and compelling to students. Kardan is talking specifically about using video games to teach aspects of Hawaiian culture and has produced a video game.

Prensky (2001, p. 29) discussed Greenfield's argument that video games lead to deep thinking and cited her view that "the process of making observations, formulating hypotheses and figuring out rules governing the behavior of a dynamic representation is basically the cognitive process of inductive discovery." Later Prensky (2005) argues that complex games are filled with choices and ethical dilemmas—more elements of deep and complex thinking enabled by video games. Squire and Jenkins

(2003) make an interesting point that the properties and processes of well designed games motivate students to turn to textbooks with a newfound attitude of a seeker of meaningful knowledge rather than merely pursuing rote learning. A strong argument can be made that learning about another culture in context and through action may lead to student engagement in deeper and more meaningful knowledge building than many traditional models that push students towards rote learning.

Much has been made of the potential benefits of informal learning through games in spaces outside of formal educational institutions. Prensky (2007, p. vii in Aldrich, Gibson & Prensky) contends that "whenever one plays a game, and whatever game one plays, learning happens constantly, whether players want it to, and are aware of it or not. And players are learning about life, which is one of the great positive consequences of all game playing. This learning takes place, continuously and simultaneously in every game, every time one plays." Learning about traditional cultures is well suited to informal settings outside of institutional learning spaces or formal settings. Foreman & Borkman (2007) point out the advantages of games for sociology students, and although they are discussing learning about contemporary society, the same principles would apply to learning about traditional society. They argue that a promising approach would be via massively multi-student online learning environments. Recently, the Australian national newspaper reported that the Australian university RMIT had purchased a virtual island in Second Life and had set up a virtual university complete with a nearby clubbing scene. It's conceivable that a virtual Torres Strait or Aboriginal island could be constructed with avatars controlled by indigenous people acting as guides for guests, showing them around the island and introducing them to the traditional culture. Kardan (2006, p. 92), when referring to his game that highlights Hawaiian traditional culture, states that "the player would gain an understanding of how the family unit functioned, what would be typical tasks performed, and what roles the adults undertook, based on their gender as well as rank in the family." Kardan argued strongly about the importance of cultural, historical, and linguistic accuracy.

Another advantage of video game creation to highlight traditional cultures is the potential for collaboration across university faculties. Faculties such as education, computer science, sociology, anthropology and business have obvious value in such a cooperative project. Collaboration with community stakeholders would be critical to the success and authenticity of the product but would also provide valuable links for future partnerships with indigenous communities. Kardan pointed out the value of close cooperation with the community and wrote that "the game design evolved over a number of meetings with such experts from the local community and the Hawaiian Studies program at the University of Hawaii. It is important for game developers to maintain flexibility in their thinking and planning and truly listen to cultural advisors" (p. 93). Likewise, James Cook University could access cultural advisors in the School of Indigenous Australian Studies and the two indigenous communities. Christie (2004) advises that we need to find appropriate ways of preserving some of the knowledge of the old people before their knowledge is lost through death. This is particularly important considering the strong oral traditions of indigenous Australians. While Christie was referring to the value of databases to preserve Aboriginal knowledge, the same objective could be achieved through video games. In a similar vein, Pumpa, Wyeld and Adkins (2006, n.p.) argue that Aboriginal people "have a real desire to preserve and pass on knowledge practices in spite of the decline in their use of traditional language, loss of ritual and passing away of elders." Pumpa et al. critique the use of databases as being flawed in that they are an excellent means of storing and retrieving vast amounts of data but do not allow for "virtual performance (traditional dance or song) and more importantly support the performance of these practices in the landscape" (n.p.). Video games on the other hand would provide an environment for the elements flagged as critical by Pumpa et al.

Disadvantages

Successfully producing video games that embed traditional culture in an engaging and beneficial manner is a complex and challenging task with many potential hurdles. These include keeping in mind that learning through video games does not suit all learners; underlying ICT structures such as game engines and databases have been described as 'Eurocentric'; game development is expensive and difficult and any inaccuracies in the game could be potentially misleading.

When Squire (2005) used Civilization III in his teaching, he found that approximately one quarter of the students elected to withdraw from the unit, choosing reading groups instead. Students cited reasons such as finding video games too hard, too complicated and, to them, uninteresting. Van Eck (2006, p. 18) cautions that "we run the risk of creating the impression that all games are good for all learners and all learning outcomes, which is categorically not the case." Creating a game highlighting Aboriginal or Torres Strait Islander culture would be an interesting addition to existing resources to archive and promote understanding of these cultures, rather than something that would appeal to all learners.

Kardan (2006, p. 93) discusses the complexity of producing video games that embed traditional culture from the perspective of someone who has completed such a task. He advises that "game development in general is laborious and complex, even without the added weight of cultural accuracy. Limiting the scope of the game and reigning in grandiose ambitions is essential." Kardan's team grappled with using an off-the-shelf game engine as a basis for creating their own game engine particularly suited to the purposes of the project. They took the more complex option of developing their own game engine. This leads to some discussion on the important point raised by researchers such as Christie (2004) and Pumpa et al. (2006) that existing ICT tools may be not culturally neutral. Pumpa et al. claim that "current technologies for the representation of this knowledge tend to embody assumptions that are based on Eurocentric scientific knowledge tradition" (n.p.). Christie argues that ICT tools "are not innocent objects. They carry with them particular culturally and historically contingent assumptions about the nature of the world, and the nature of knowledge; what it is, and how it can be preserved and renewed (p. 1). Taking these cautions into account, Kardan's option to create a purpose built game engine may be warranted.

Conclusion

Weighing up the advantages of undertaking this project against the many challenges, it seems that potential benefits would make it worthwhile to optimistically proceed with the game development and the associated research on its effectiveness, using a case study approach as outlined by Yin (2006) to gather and analyse data. Early adopters of this approach, including the Hawaiian team led by Kardan, have experienced some success. Prensky (2006, p. ix) maintains that "we are just beginning to understand and document, through a blend of IT and social research, how games and simulation elements can take advantage of the global network infrastructure." In addition, Khaled, Barr, Fischer and Noble (2006, p. 213) explain that "we have witnessed a surge in the development of games that are aimed at changing people's attitudes towards particular topics, instead of being strictly entertainment oriented." The combination of the potential of global networking and the promise of influencing attitudes in a positive way, while developing specific knowledge of indigenous cultures, is compelling. With the development of sophisticated technology and a will to preserve, highlight and educate others about traditional Aboriginal and Torres Strait Islander culture, this project has the potential to be successful and beneficial. Research highlights the current lack of game-based models associated with fostering knowledge of, and interaction with, traditional cultures and supports the view that more development is needed in this area.

...

PART 6

Knowledge Transfer

WITH LYN COURTNEY

Equity and Knowledge Transfer

The success of knowledge transfer between Western and Eastern countries, between developed and developing countries or other combinations has emerged as an important equity area during the era of globalization. International knowledge transfer facilitated by ICT has the potential to improve conditions for people in all countries since knowledge transfer is a two-way process. Part of knowledge transfer involves access to information. Flor (2001, p. 1) argued that "since the 'information society' concept was introduced in the seventies, the correlation between access to information and poverty has been widely acknowledged. The main propositions given were as follows: information leads to resources; information leads to opportunities that generate resources; access to information leads to access to opportunities that generate resources." Access to codified information is only part of the nature of knowledge and technology transfer. Ernst and Lundvall (2004) considered ICT to be a critical factor in the sharing of tacit knowledge and the development of new shared tacit knowledge. They pointed out that "IT, in the context of globalization, speeds up the rate of economic change and that, as a result, the need for rapid learning of tacit as well as codified knowledge has dramatically increased" (p. 275). The notion of knowledge transfer is different from, but related to, the concept of 'digital divide,' since lack of access to ICT resources and training can have a negative effect on knowledge transfer.

Knowledge transfer can occur without access to Information Communication Technology and can involve any area of human endeavor. For example, one of the cases described in the following chapter involves health education and the sharing of knowledge between Australia and China about preventing and treating depression.

Putting Knowledge Transfer into Context

Knowledge is considered the single most important driving force in acquiring, retaining and sustaining economic growth (Grant, 1996; Knight, 2003) and competitive market advantage (Leonard-Barton, 1995; Smith, 2001) in an Information Communication Technology (ICT) society. With the advent of the 'information society,' there has been a knowledge revolution which demands imaginative and intuitive directors to manage knowledge and convert this knowledge into useful products (Goffee & Jones, 2000). However, there has been little research conducted on international knowledge transfer (Li-Hua, 2003, 2007).

Hetman (1973) identified six main criteria for knowledge transfer and technology assessment studies: technology, economy, society, the individual, the environment and the value system. Similarly, Samli's (1985) model identified geography, culture, economy, business, people and government as critical factors in achieving effective knowledge and technology transfer. It is clear from the literature that effective knowledge transfer relies on successful communication between people from diverse cultures, often with differing value systems operating in business and economic environments that may be highly constrained by government (Grant, 1996; Li-Hua, 2007; Spender, 1996).

Universities have expanded their traditional role from teaching, learning and research (e.g., creating intellectual capital) to providing bi-directional knowledge transfer between partner universities, industry and government. Successful tertiary knowledge transfer has been identified as a relatively little known resource that contributes substantially to the development of economic competitiveness and the creation of wealth in both developed and developing countries. However, because of limited empirical investigation of knowledge transfer in higher education there is limited understanding of how best to achieve successful knowledge transfer in this area (Li-Hua, 2007).

This chapter will attempt to put knowledge transfer into context by first defining the terminology and then providing some insights into knowledge transfer in developing countries. Next, this chapter will briefly outline Australia's and China's geographic and economic landscapes, which provides the context for the next section, knowledge transfer between Australia and China. A brief overview of tertiary education knowledge transfer will conclude with a review of a study currently being undertaken at James Cook University into international bi-directional academic and research knowledge transfer between Australia and China. The concluding section will examine knowledge transfer between the United States and China.

Defining the Key Knowledge Transfer Concepts

Knowledge

As the world transcended into the 'information age,' organizational and psychological theorists began to explore the realm of intangible corporate and tertiary education assets, such as the knowledge and expertise of employees (Spender, 1996). Similar to intangible organizational assets, such as corporate reputation and brand identity, knowledge is now considered an organizational asset (Pascarella, 1997). Blackler (1995) proposed that knowledge is complex and multifaceted. However, knowledge is a highly contentious construct (e.g., Spender, 1996); therefore, before knowledge transfer can be adequately discussed, some defining of the construct of knowledge is necessary, especially considering the recent shift to describing our times as the 'knowledge age' rather than the 'information age' and the 'knowledge society' rather than the 'information society.'

Alavi and Leidner (1999) suggested that knowledge is a "justified personal belief that increases an individual's capacity to take effective action" (p. 5). Knowledge refers to what is known by perceptual experience and reasoning. There are important differences between data, information and knowledge, which are not interchangeable terms. For example, 1234567.89 is data; "Your bank balance has jumped 87% to $1234567.89" is information; "Nobody owes me that much money" is knowledge; and "I'd better talk to the bank before I spend it because of what has happened to other people" is wisdom (Interoperability Clearinghouse Glossary of Terms, 2008). Knowledge is broader than intellectual capacity because it is dynamic and is the product of action and interactions between individuals (Odell & Grayson, 1998). Actionable knowledge refers to knowledge that produces action or innovation, such as making people aware of possibilities and how to make possibilities into realities (Buckley & Carter, 2000). This type of knowledge has been identified as the single most important factor, and a strategically essential resource that needs to be carefully managed in order to ensure survival of organizations in the competitive knowledge society (Grant, 1996, Nonaka, 1994).

Knowledge can be broadly categorized into explicit knowledge and tacit knowledge (Nonaka, 1994; Polanyi, 1968). Both types of knowledge are important, as there is a complex interaction between explicit and tacit knowledge. The interaction of explicit and tacit knowledge is central to Nonaka and Takeuchi's (1995) theory of organizational knowledge, which expands the view of production and resource based assets to include knowledge based assets.

Explicit knowledge

Explicit knowledge is formal—written or codified (O'Dell & Grayson, 1995). According to Rahim and Golembiewski (2005) explicit knowledge can then be divided into endogenous and exogenous explicit knowledge. Endogenous explicit knowledge refers to knowledge such as operations manuals, codes of practice and databases, whereas exogenous explicit knowledge refers to things like competitors' products or patents (O'Dell, & Grayson, 1998; Rahim & Golembiewski, 2005). Explicit knowledge transfer is direct, clearly specified and relatively easy, as it requires no conversion. "Explicit knowledge is technical and requires a level of academic knowledge or understanding that is gained through formal education, or structured study" (Smith, 2001, p. 315).

Tacit knowledge

Tacit knowledge is informal—unwritten, non-linguistical or uncodified knowledge (O'Dell & Grayson, 1998). It has also been described as "disembodied know-how that is acquired in the informal take-up of learned behaviour and procedures" (Howells, 1995, p. 2). Know-how encompasses things such as product design, quality issues, and distribution expertise. According to Rahim and Golembiewski (2005) tacit knowledge can also be divided into endogenous and exogenous tacit knowledge. Endogenous tacit knowledge refers to knowledge that is in employees' heads, such as their skills and expertise, whereas exogenous tacit knowledge refers to things like suppliers' or customers' ideas. Tacit knowledge is internalized by the knower, not articulated, experiential, transitory and difficult to document (O'Dell & Grayson, 1998).

Nonaka and colleagues (1991, 1995) applied the concept of tacit knowledge to knowledge management based on Polanyi's (1968) concept of tacit knowledge being a process. Nonaka and Takeuchi (1995) created a model of knowledge creation. They suggested that tacit knowledge is transferred through a learning process and does not have to be converted to language, although language is one way of transmitting tacit knowledge. Von Krogh and Roos (1995) argued convincingly that tacit knowledge is a characteristic of an individual alone. However, Baumard (1999) demonstrated that there is evidence to support the notion that individuals may possess knowledge that they do not realize they have learned and do not know they possess. In contrast to von Krogh and Roos (1995), he argued that tacit knowledge can be an attribute of an individual but also of a group that works together and in organizations that demonstrate shared knowledge.

In their critical examination of the literature on tacit knowledge, McAdam, Mason and McCrory (2007) surmised that "an improved understanding of tacit knowledge is needed to underpin and further develop the knowledge management discourse" (p. 43). Researchers have asserted that tacit knowledge is the most important strategic resource of an organization to sustain competitiveness (Grant, 1996; Nonaka, 1994) and to facilitate learning (Lam, 2000). It appears that one of the keys to knowledge transfer is to not only transfer explicit knowledge, but to find a way to make tacit knowledge explicit knowledge so that it is transferable as well. McInerney (2002) suggested that the best way to do this is for organizations to create a "knowledge culture" (p. 1014) that fosters bi-directional learning, creative innovation and sharing of knowledge.

Finally, there is articulable tacit knowledge, which refers to knowledge that can be "articulated for practical and competitive reasons" within an organization (Richards & Busch, p. 1). Organizational research stresses the importance of tacit knowledge for organizational effectiveness (Goldring et al., 2008) in which shared frames of reference can be applied to determine "the way work is really done" (Smith, 2001, p. 316).

In 2000, Egbu developed a framework for managing knowledge in which people, content, culture, process, infrastructure and technology were identified as critical factors and the complex component of tacit knowledge identified as a point of knowledge blockage. Knowledge blockage results because tacit knowledge has to be transferred through personal human interactions (Tsang, 1995), or exchanged only at the individual level (Fallah & Ibrahim, 2004). This requires a much higher intensity and quality of communication, which is time-consuming and harder to achieve. McInerney (2002) suggested that the transfer of tacit knowledge into explicit knowledge formats is both challenging and controversial.

Knowledge transfer

Knowledge transfer refers to a process through which an individual or organization is "affected by the experience of another" (Argote & Ingram, 2000, p. 151) or the ability to create and distribute knowledge to the mutual benefit of all stakeholders. A "healthy system of knowledge transfer should demonstrate considerable diversity in knowledge transfer approaches and activities, both within and across institutions" (Department of Education, Science and Training [DEST], 2006). DEST (2006) has suggested that the literature on the concept of knowledge transfer denotes a one-way flow of knowledge; however, this chapter has adopted the understanding that knowledge transfer is a two-way, or bi-directional, negotiated flow of knowledge for the mutual benefit of the stakeholders.

Nonaka (1994) forwarded four different modes of knowledge transfer: socialization, externalization, internalization and combination. Socialization is said to occur when an employee acquires and exchanges knowledge through discussions with other employees of the organization. Externalization occurs when an employee of one organization writes down information gathered by attending a meeting with employees from another organization. Internalization occurs after an employee in one organization reads about events of the organization and mentally combines it with previous experience. Harman and Brelade (2003) cautioned that power imbalances may occur when knowledge is transferred within or between organizations and nations.

In order to achieve bi-directional knowledge transfer, three elements must be considered: the demand side, the supply side and engagement (Bishop, 2006, n.p.; DEST, 2006):

» The demand side: one or more stakeholders who have specific needs and characteristics that research and research based capacities could help. It is critical "to create incentives for knowledge transfer on the supply side of the equation" (DEST, 2006, p. 72).

» The supply side: one or more stakeholders must have particular areas of strength that are recognized and able to be transferred to the other stakeholder(s). To be responsive to the demand side, both stakeholders must have in place "supporting cultures, structures and systems and need to have the requisite capacity to adopt a strategic approach and to dedicate resources to effective knowledge transfer strategies and priorities" (DEST, 2006, p. 72).

» Engagement: one or more stakeholders need to have effective mechanisms for both the demand and supply sides in order to meet and understand each other, thereby being able to develop mutual strategies.

Knowledge transfer is comprised of "multiple inter-connected and overlapping processes" (DEST, 2006, p. x). Recently, Partners (2005) identified four knowledge transfer processes: knowledge diffusion, knowledge production, knowledge relationships and knowledge engagement. Knowledge diffusion refers to making knowledge accessible to users through media, such as through publications, conferences, professional education. Knowledge production refers to the sale of knowledge products, such as patents, licensing multimedia products. A knowledge relationship consists of the sale of knowledge services, such as consulting, contract research and education and training contracts. Finally, knowledge engagement refers to engagement to achieve mutually beneficial outcomes, such as alliances aimed at achieving mutually beneficial goals (e.g., exchange of research findings) (Partners, 2005).

Knowledge spillover

Knowledge spillover is the unintended transfer of knowledge to unintended recipients and occurs without the consent of the owner of the knowledge (Smith, 1999). Knowledge spillover in technological clusters is suspected to be the source of the increased rate of innovation (Fallah & Ibrahim, 2004). Jaffe (1989) began measuring the effects of knowledge spillover and innovation. Jaffe, Trajtenberg, and Fogarty's (2000) research into knowledge spillover resulted in a proposed conceptual model of knowledge transfer and knowledge spillage that contributes to increased innovative output of technological clusters. Whereas tacit knowledge is often thought to occur primarily at the individual level, it is clear that knowledge spillover occurs at an individual, enterprise and even national level (Fallah & Ibrahim (2004).

Knowledge management

Despite some positive aspects of knowledge spillover, sometimes knowledge spillover is not desirable and may be considered problematic. Therefore, the world is moving beyond information management and towards knowledge management in order to ensure competitive advantage (Selamat et al., 2006). Valuable knowledge can be lost unless it is managed effectively, and tacit knowledge is particularly vulnerable due to organizational downsizing, which results in employment terminations (Smith, 2001). There is little consensus in the literature regarding a comprehensive definition of knowledge management (Bhatt, 2001) despite it being attributed with improving work processes and creating value for organizations (Selamat, Abdullah & Paul, 2006). However, Schulze (1998) proposed that knowledge management consists of organizational learning and memory, including information sharing and collaboration. Implied in the definition is that there is a systematic and organizationally specified process for "acquiring, organizing, sustaining, applying, sharing and renewing both explicit and tacit knowledge that enhances the organizations performance, creating value" (Alavi & Leidner, 2001, p. 107).

Alavi and Leidner (2001) suggested that in order for different individuals to arrive at the same understanding of knowledge, they must share a history or context. In this way, technology makes information both relevant and accessible, regardless of its location, and fosters an environment and culture in which knowledge can evolve (Davenport & Prusak, 1998; Selamat et al., 2006). However, when approaching issues of knowledge transfer between First World Countries and Second and Third World countries, issues of dissimilar country histories and contexts can be the pivotal barrier to effective knowledge transfer.

Knowledge Transfer in Developing Countries

After World War II, many former European colonies in Africa and Asia emerged as new nations in need of national infrastructure for the development, manufacture and exportation of capital goods (Akubue, 2002). Technologically advanced Western countries provided the technical experts, infrastructure and modern technology to industrialize developing countries. Globalization offered unique opportunities for developing countries to acquire advanced technology and knowledge, thereby leveling the economic playing field (Li-Hua, 2007).

However, despite the major injections of technology, many developing countries today remain economically bankrupt. It was previously assumed that technology which works well in the country of origin would be equally effective when transplanted into another country. However, technology transfer failures in developing countries have revealed that effective knowledge transfer consists of more than the mere acquisition of physical assets (Akubue, 2002). Samli (1985) agued that successful knowledge transfer is related to its appropriateness in meeting the needs and conditions prevailing in the receiving country. This requires understanding that the country's language and culture as effective knowledge transfer relies on the successful communication, or transfer, of ideas.

Tertiary Education Knowledge Transfer

Universities, or knowledge institutions, are currently one of the underexploited economic resources because those outside of tertiary education often do not know how to access this knowledge and those inside of tertiary education often do not know how to connect to the outside world (Li-Hua, 2007). The report Tertiary Education for the Knowledge Society (Organization for Economic Cooperation and Development [OECD], 2008) reviewed tertiary education policy from 2004 to 2008 in collaboration with 24 countries worldwide. This report highlighted the important contributions of tertiary education to economic competitiveness in the increasingly knowledge driven global economy. This report also noted that tertiary education is more internationalized, which consists of intensive knowledge sharing between institutions, academics, students and industry. This process has fostered international collaborative research activities, which contribute to the economic and social goals of participating countries, such as business and community development which contributes to international competitiveness. In addition, tertiary institutions often provide consultancy services to industry and governments (OECD, 2008).

Australia enjoys a strong scientific and ICT knowledge base (Commonwealth of Australia, 2005). The Australian government is committed to investment in science and innovation, which includes generating new ideas and undertaking research to ensure a secure future for Australia. For example, the Australian government has put aside approximately $46 million in funding to promote collaboration between universities and industry (Bishop, 2006). Furthermore, the Australian Research Council (ARC) has allocated approximately $260 million in ARC Linkage programs designed to develop new knowledge or innovation in an environment of partnerships between universities and industry (Bishop, 2006). The Backing Australia's Ability report (Commonwealth of Australia, 2005) highlights the advances Australia has made in ICT and supports the creation of the world class ICT Centres of Excellence in order to continue to grow Australia's international profile as a strong competitor and contributor of ICT. Australia's global position in science, ICT, higher education and research provides enormous opportunities for Australian universities to engage in academic and research knowledge transfer partnerships internationally, particularly in the Asian Pacific region.

Research institutions in Australia are realizing that they have an important role in contributing to the global economy, and they have been involved in interesting initiatives to not only ensure that their graduates are prepared to compete in an international marketplace, but that world class research is undertaken in such a way as to capitalize on their science and ICT strengths (DEST, 2006). Sharing knowledge with international academic, business and government stakeholders through research and development collaboration provides a potential source of income for universities and also increases both the quantity and quality of research undertaken. Furthermore, research institutions are recognizing the need for them to engage in a more active role in managing knowledge and transferring knowledge to ensure bi-directional benefits for all stakeholders (DEST, 2006). In this way higher education and research activities can attempt to meet the needs of different societies by improving international collaboration by supporting effective knowledge transfer activities.

Tertiary knowledge transfer is recognized to produce an assortment of benefits depending on the stakeholders. For example, knowledge transfer between academic institutions is directed towards enhancing "material, human, social and environmental wellbeing" (DEST, 2006, p. viii). Conversely, knowledge transfer for commercial benefit is directed "towards enhancing the success of commercial enterprises" (DEST, 2006, p. viii).

Some cultural barriers to knowledge transfer have previously been mentioned; however, research findings by Spendlove (2005) highlight an additional barrier to effective academic knowledge transfer—a stereotypical perception that universities are 'Ivory Towers,' complete with specialized language and intellectual jargon. One way of overcoming this barrier would be to ensure that common language is used in the process of knowledge transfer. In addition, this study revealed that the lack of people skills was a threat to knowledge transfer and innovation.

Using the authors' institution as an example, James Cook University (JCU) is actively involved in expanding its campuses in offshore facilities and partnerships as it continues to expand as a multi-campus university. JCU has established numerous partners in China and has agreements with a number of Chinese universities. Therefore, research in knowledge transfer is a critical step in securing successful multinational educational and research ventures.

Knowledge Transfer: A Case Study

To understand the environment in which knowledge transfer occurs, it is essential to begin to understand the similarities and differences of the people and countries in which knowledge transfer is expected to occur. Therefore, this section will briefly compare China with Australia and the Chinese with the Australians.

China is the largest country in the world with a population in mid-2007 estimated at 1.3 billion, about one-fifth of the world's population (Central Intelligence Agency [CIA], 2008). While there has been a program of economic development, many Chinese live in poverty as farmers or herders because keeping the large population of China fed is a major challenge. In contrast, Australia is the sixth largest country in the world with a population in mid-2007 estimated at only 20.4 million (CIA, 2008). Australia is a prosperous country, and over 70% of Australians live in large cities with farming also being a major industry. The life expectancy in China (72.88 year for males and 71.13 years for females) is almost eight years lower than Australia (80.62 years for males and 77.75 years for females) (CIA, 2008).

China is comprised of 22 provinces, four autonomous regions, four municipalities and two special administrative regions of Hong Kong and Macao (Lunn, Lalic, Smith & Taylor, 2006), whereas Australia is comprised of six states and two territories (CIA, 2008). The political organization that holds the real power in China is the Communist Party, with membership of approximately 70 million (Lunn, et al., 2006), whereas Australia is a Constitutional Monarchy (Commonwealth of Australia, 2008). Unlike Australia, China has a history of political and social turbulence, which has included civil war and long periods of authoritarian rule (Lunn, et al., 2006). Siach (2004) highlights the complexity of Chinese polity and cautions that "the rules of the game" (p. 95) are poorly developed, opaque and subject to change.

China began the process of economic reform in the late 1970s with its 'open door' policy, which allows foreign trade and investment. In 2001, China joined the World Trade Organization, and Sachs (2003) predicted that China has the potential to have a larger economy than the U.S, by 2041 as long as China continues to foster economic policy that is conducive to growth. Both China and Australia have enjoyed strong economic growth for a number of years. China's economic growth has exceeded the economic growth of the United Kingdom (UK), the United States (U.S.) and Japan since 1990 (Lunn, et al., 2006). The Gross Domestic Product (GDP) of China is $7.043 trillion (2007 estimate), whereas the GDP of Australia is $766.8 billion (2007 estimate) with the real GDP growth rate of China being 11.4% (2007 estimate) compared with Australia's 4% GDP growth rate (2007 estimate) (CIA, 2008). The estimated unemployment rates for China and Australia are very similar, with China reporting 4% unemployment (2007 estimate) compared with 4.4% unemployment in Australia (2007 estimate) (CIA, 2008). Excluding goods exports from Hong Kong, China's goods exportation has grown rapidly in recent years, and they are now the world's third largest goods exporter after the U.S. and Europe with nearly 10% of the world's total exports and 8% of the world's imports (Lunn et al., 2006). It is noted, however, that the Chinese business environment is fraught with ambiguities that are poorly understood, which makes knowledge transfer between China and the global economy difficult (Li-Hua, 2007). It should be noted that Australia is a very small market in comparison with the global marketplace (Partners, 2006).

China is not as technologically advanced as Australia in mobile phone or Internet use. China had 461.1 million mobile phones in 2006. Once children 14 and under were removed from the overall population, this computes to 0.44 mobile phones per person aged 15 and over compared with 9.94 million mobile phones in Australia in 2006, which computes to 1.20 mobile phones per person aged 15 and over. Furthermore, Internet users are more prevalent in Australia, with 92% of the population 15 years of age and older being Internet users in 2006 compared with 15% of Chinese aged 15 and over being Internet users in 2006 (CIA, 2008).

Of critical importance to knowledge transfer is freedom of expression and access to information. Successful knowledge transfer does not necessarily require face-to-face communication. In today's Internet society, one of the ways to expand access to information is through electronic knowledge transfer. According to Roberts (2003), before electronic communication can successfully contribute to knowledge transfer, trust must be established between the stakeholders.

Project and aims

There is extensive evidence that technology and knowledge transfer between Australia and China is not always successful, with misunderstandings and cultural differences being mentioned as barriers to successful knowledge transfer (Li-Hua, 2007). Li-Hua (2003) argued that without knowledge transfer, technology transfer does not occur. However, there has been little empirical investigation of knowledge transfer in tertiary education. Li-Hua (2007) conducted a review of the literature with respect to education collaboration between the United Kingdom (UK) and China, and also conducted a study designed to identify the elements required for successful tertiary knowledge transfer. There was a consensus from the interviews with university upper management from the UK and China that internationalization is a high priority and international academic collaboration was seen as mutually beneficial. Chinese university heads complained that they often lost their students to foreign universities, which was undesirable, and there is interest in adopting a "double-campus" model so that cooperative education and research projects can be carried out within the Chinese educational system (Li-Hua, 2007).

The current research project aimed to investigate the mechanisms of, and barriers to, knowledge transfer between Australia and China in tertiary education. Of particular interest is how academic and research knowledge is transferred and shared; however, business knowledge transfer was not excluded, as it may provide additional insights and perspectives into the commercialization of research findings.

Methodology

Participants

Twenty Australian and Chinese individuals from academic, research and business were identified as having been engaged in knowledge transfer between Australia and China. Ten Australian and one Chinese academic accepted the invitation to participate in individual interviews investigating their knowledge transfer experiences and the potential of commercialization of research findings. Of these 11 participants, four were females and seven were males.

Materials and procedure

The individual focused interviews (Yin, 1994) were conducted in person or by telephone and digitally recorded. The interviews lasted between 45 minutes and an hour and with the author using a structured interview style, covering ten questions listed below as well as open-ended questions to allow the participants ample opportunity to elaborate on the intricacies of their knowledge transfer experiences:

1. How is academic knowledge transferred between key academic/research staff in Australia and their Chinese counterparts?
2. How is knowledge transferred between individuals, businesses, companies and subsidiaries?
3. What type of knowledge do you value for sharing?
4. How much does the capacity (e.g., the relationship between a full campus and a satellite campus or a large company with a small Chinese subsidiary) of the Chinese partner have on knowledge transfer?
5. What have you identified as barriers for knowledge sharing (e.g., cultural differences, values)?
6. What are the key factors that facilitate knowledge transfer (e.g., relationship, language)?

7. Does the type of relationship (e.g., equal partners, small company vs. large organization) have an impact, either positive or negative, on knowledge transfer?
8. Do you think there is potential for commercialization of research findings between Australia and China?
9. Do you know of any examples of commercialization of research findings?
10. Do you consider that ICT has an important role to play in knowledge transfer? If yes, please explain why ICT has an important role in knowledge transfer?

The data were transcribed verbatim and subjected to content analysis whereby the data were initially open coded (Strauss & Corbin, 1998) and categories identified, using comparison within and between interviews (Boeije, 2002; Tesch, 1990). In addition to open coding, appropriate forms of sociolinguistic analysis (Gee, 2005; Johnstone, 2002) were applied to the data to focus on the wordings participants used to describe the processes they go through when engaging in knowledge transfer (e.g., verbs denoting emotional or reasoning processes).

This section reports primarily the analysis of responses to Question 1. "How is academic knowledge transferred between key academic/research staff in Australia and their Chinese counterparts?" and Question 10. "Do you consider that ICT has an important role to play in knowledge transfer? If yes, please explain why ICT has an important role in knowledge transfer?"

Results and discussion

Preliminary content analysis was undertaken for the two questions that are the focus of this chapter. These results will be presented and integrated with discussion, in separate subsections.

How is academic knowledge transferred between academic/research staff in Australia and their Chinese counterparts?

Preliminary content analysis results for the question "How is academic knowledge transferred between academic/research staff in Australia and their Chinese counterparts?" provided a range of responses, which included knowledge transfer between Australian and Chinese academics working collaboratively to develop research partnerships and training, scholarships and academic exchange programs, joint degree programs, and collaboration on research publications.

It was acknowledged by one participant that China did not wish to be treated like a developing country but had "aspirations to be an education hub of the world" and that "they are putting in vast amounts of infrastructure in order to accomplish this goal." However, while China is developing this educational and research infrastructure, a participant noted,

They send many of their scholars to Australia to be trained in an English research environment...because they thought English was the language of scholarship and that was essential if they were to be taking their rightful place in the world of research and innovation.

One aspect of research collaboration and publication addresses the aspiration of publishing in both English and Chinese. Part of this collaboration process involves

doing a lot of editing, so that it would come out in fluid English...if it is going to be finally written in Chinese, the Chinese side cleans up the Chinese form and if it is finally written in English, we clean up their English.

This involvement goes beyond literacy issues but encompasses writing in culturally appropriate ways so that the essence of the message in research publications has integrity. Another participant noted,

If there are language barriers, then it is quite different for someone in China to sit down and pour over the English and laboriously translate and make sure they get every innuendo. This is going to work with more precision than conversations.

A hindrance to knowledge transfer was noted by three participants—that the transfer had been primarily a one-way transfer from Australia to China. This is contradictory to the ideal of creating a knowledge culture with bi-directional knowledge sharing (McInerney, 2002). For example, these statements from three participants,

It is my belief with Chinese academics, and Chinese people, that they actually don't interact verbally a great deal. There is definitely a cultural thing where they don't share knowledge. They soak knowledge up but don't give knowledge out. Free sharing of knowledge is not part of their culture.

The Chinese have a very transactional view of all this.... they would like to see as much transferred in their direction and as little the other way. What we have here, from the Chinese lecturers, they will not share what they know. Like it is really hard to understand what they are even doing in the classrooms in terms of teaching.... everything is kept to themselves as if they personally own it.

Conversely, it was also noted that "Chinese universities have been working very closely with Australian universities to provide courses and things like traditional Chinese medicine, acupuncture." It was also noted that

Australian universities have done very well in recruiting some very bright people, particularly in the science area, from China, working in Australian universities. Although often when you look at those people you actually find that they came over here as undergraduates to go into graduate degrees here and then doctorate degrees and on to be academics.

However, this same participant said that there is currently a reversal of this process and gave solar technology and the renewable energy sector of education as an example,

You are seeing of the quite senior Chinese academics going back to China....and that places like the University of ***** have lost some really significant academics to China because China has been willing to put in the infrastructure and research resources that those academics needed.

Another form of knowledge transfer mentioned by most of the participants was with regard to intellectual property (IP), such as course materials. Some participants did not feel that course materials constituted real IP, for example:

I think there is so little intellectual property in teaching material in subject courses and I don't believe that is what they are really buying or what they are interested in.

Alternatively, other participants suggested that a problem encountered in teaching was that they actually "had to give the IP over to the university" (meaning the Chinese universities).

Several participants talked about 'personal IP,' which supports other researchers' conclusions about tacit knowledge being a characteristic of the individual (e.g., von Krogh & Roos, 1995). One participant said that where IP really starts getting important is when it's personal IP, or tacit knowledge, and how that knowledge is shared. For example,

The individual's intellectual property that they have put into that compartment of 'knowledge' that they have created around topics. And that sharing only comes by close association with that person, how they teach and what it is. The detailed contact that they are actually delivering plus the pedagogy and the processes that they use to impart learning. And that is a little bit more passive and takes a lot more time for that sort of transfer to occur.

Finally, one participant stated that knowledge transfer between China and Australia is further compromised "because they also want to filter the IP."

One way in which to enhance knowledge transfer would be to establish trust between Australian and Chinese academics and researchers. More than 80% of the participants specifically stated that establishing "trust" and "mutual understanding" was essential to forming strong relationships where authentic knowledge transfer can occur. For example, the following comments were made by three participants about the importance building relationships based on trust before bi-directional knowledge transfer can be expected:

> There is still probably a degree of lack of trust and lack of empathy with what the two partners are actually trying to achieve.
>
> This is going to require building trust and the Chinese are usually deeply suspicious of whether this is just another university road show and whether you will come back. I would find until you have actually been back three or four times, it is unlikely that you are going to get anything productive happening.
>
> China is very different. It almost sounds cliché, but it is about building up relationships and you are going to have to build up relationships over a long period of time before you do get the trust and before you do get true knowledge transfer, before the Chinese, I think will trust to give their knowledge.

In accordance with the literature about the importance of understanding cultural differences (e.g., Li-Hua, 2007; Samli, 1985; Spender, 1996), the following comments from three participants reflect how important cultural understanding is to knowledge transfer:

> The very active use of facilitators who have that cross cultural understanding is essential, on both sides, and I guess I think a bit of humility on behalf of some Australians…. and a better appreciation of the spoken magnitude of China, the history, the impressive record of scholarship and the fact that Chinese are very proud people and that it is really important that we acknowledge that.
>
> I mean if you want to do business in China, you have to understand how China works. Well I think the first thing is to facilitate knowledge transfer you need to have an understanding of the people you are dealing with. That doesn't matter whether they are Chinese academics or German academics, it is a matter of understanding the cultural constraints they work under.

It is clear from the comments of the participants in this study that knowledge transfer is comprised of complex interactions between people and institutions. When aspiring to achieve successful transfer between academics and researchers in two countries, it is imperative to understand the diverse values, cultures, socioeconomic and government systems of the two countries for knowledge sharing to transpire.

These findings support previous research (e.g., Grant, 1996; Li-Hua, 2007; Smith, 2001).

Do you consider that ICT has an important role to play in knowledge transfer? If yes, please explain why ICT has an important role in knowledge transfer?

Content analysis of the responses to Question 10 on the importance and role of ICT in knowledge transfer revealed the themes of:

a) Importance of ICT
b) Types of ICT, their perceived importance and frequency of use
c) Challenges associated with ICT use

Importance of ICT

It is not often that any research question evokes complete consensus in a group of diverse participants. Although this study included only 11 participants, they were key academics, spread across many Australian universities and locations. The question "Do you consider that ICT has an important role to play in knowledge transfer?" received an overwhelmingly positive response. The importance of Information Communication Technology (ICT) to enhance knowledge transfer and knowledge management and to contribute to developing economies has been long recognized. The 2000 Okinawa Summit consisting of G7 and G8 nations recorded that ICT was one of the most potent forces in shaping this century and described it as being a "vital engine" for growth of the world economy. One participant agreed with the importance of ICT but expressed some reservations:

> Well ICT, obviously brings up the speed of communication incredibly—and that is good, but I'm not convinced that ICT is the way of necessarily overcoming that cultural barrier to affect communication. And, in fact, in some ways, because it replaces some of the other mechanisms of communication, I wonder whether it may exacerbate the cultural barriers.

The same participant felt that ICT based communication should come after initial face-to-face development of relationships:

> I think all of those ways can work very well when you have got the relationship to a certain stage. But in the early stages, I think you really need face to face communication, there is no substitute.

In answering the second part of the question, participants described the types of ICT commonly used in their relationships with Chinese academics that fostered knowledge transfer and provided their perceptions on the frequency of use and their order of importance. The most commonly used form of ICT was e-mail, followed by sharing of information through websites, SMS, instant messaging, phone (including VOIP), Skype voice, Skype video and full videoconferencing through standalone videoconferencing units such as Tanberg systems; FTP (file transfer protocol) was also used by some participants. Web based materials were seen as important for sharing codified knowledge, whereas e-mail, messaging, voice and video were seen as essential for the sharing of tacit knowledge. Ernst and Lundvall (2004) recognized the importance of ICT as being a conduit to the sharing of tacit knowledge rather than a substitute for it. They argued that with ICT use "the emphasis is upon its potential to reinforce human interaction and interactive learning. Here the focus is not upon its ability to substitute for tacit knowledge but rather on its ability to support and mobilize it. E-mail systems connecting agents sharing common local codes and frameworks of understanding can have that effect, and broaden access to data and information among employees and can further the development of common perspectives and objectives…multimedia exchange may be helpful in transferring elements of tacit knowledge by using combinations of voice and pictures interactively" (p. 274).

Types of ICT use

E-mail and messaging was singled out as being convenient due to ease of use and speed. Examples include:

> Yes, I do since ICT can give rapid and convenient service for the knowledge transfer, especially for the two distant countries

> Email is our primary means of day to day communication

Even with simple forms of communication such as e-mail, there was some reported problems with blocking or filtering in China.

> So if I go into China, I can't access my……webmail. I can't access the website, so I have to use my Gmail account or go through some other roundabout mechanism. Sometimes we can trick the computers by just putting in the IP address rather than the conventional way and you can get around it but 99% of the time it is just a real pain. And so not being able to use it really reinforces how important it is to keep the communication going.

Another participant related that

> I have been working with a Chinese postgraduate student who has helped me break down some barriers and in communicating with some Chinese universities, she has told me that sometimes spam filters of Chinese universities block emails from the West and sometimes it's very hard to get

emails through, unless you use Chinese characters, which we have been doing.

After e-mail, the Internet—webpages and learning management systems such as Blackboard and WebCT—were thought to be most valuable, with some participants rating it higher than e-mail.

The Internet is the most valuable; the World Wide Web is the most powerful tool for disseminating knowledge.

I was involved in the National project on mental health and depression initiative into China with the Chinese Psychiatrists Association and so we took the project work from Australia and translated it into Chinese and then they launched it in China. That project, the whole point of it was to get the flexibility offered by the project website into the Chinese context. And the project team was happy because they got to expand what they think is a really good idea to a really important market. And the Chinese thought it was great because they didn't have to reinvent the wheel. So ICT was fundamental. The basic stuff—email and being able to load stuff onto the website that they can download is really important.

We often use the web and we do use SMS.

Email, telephone and sometimes Internet chat.

Access to web-based databases is very important

The program Skype was mentioned by several participants as being a valuable tool for knowledge transfer. Skype can be used in a variety of ways depending on the bandwidth and filtering. It can be used solely for instant messaging or it can support purely audio communication or can be used for webbased videoconferencing if the computers are equipped with webcams.

Yes, I regularly use Skype, so the person who was working on the mental health project who was based in Beijing communicated via Skype all the time. She didn't have the video option, so it was audio with her. Another person I work with in Hong Kong uses video from home and I also use it from home since I don't have a webcam at work, but I've got my MAC at home so I use the face to face there.

Yeah, well, we certainly did a bit of that (videoconferencing) there are the Internet ones like Skype.

Participants discussed their current use of e-mail, websites, messaging, Skype and telephone as commonly used tools but also expressed their appreciation of the potential of full videoconferencing to enable knowledge transfer. In the past full videoconferencing facilities were very expensive since they relied on ISDN links. This necessitated the installation of dedicated lines with associated high monthly rental charges by telephone companies along with a hefty fee per hour for international link-ups. More recent videoconferencing facilities make use of university broadband access and connect via IP. Since this is an emerging form of ICT, there was little evidence amongst the participants of extensive use but there was an awareness of its potential. Some indicative responses from participants include:

> Not only would it be an advantage for staff management meetings and transfer there, it would also help in terms of having special events in Australia or China videoconferenced.

> I'm trying to—we have done some videoconferencing trials on a small scale. The University of…actually teaches a Masters degree thereby videoconferencing, so it is possible. How good it is, I don't know.

> Well, you know there are a number of universities doing extensive videoconferencing into China from other countries. I mean I know many of them have trained thousands of MBA's using it, how effective it is, I don't know.

Challenges to ICT use

The participants saw the four main challenges to ICT use as:

a) Unequal access to ICT hardware and broadband connections in China
b) Blocking and filtering
c) Concerns about intellectual property
d) Knowledge transfer via ICT needs to be seen as a two-way process

Many participants highlighted the uneven development in China and viewed the variable distribution of access and resources to be the major impediment to ICT mediated knowledge transfer. The following comments from different participants illustrate the difference between the centres of urban development and the rural majority.

> Well of course ICT has an important role to play…I went to some of the IT universities in China and they make our stuff look like kindergarten— the stuff they've got. So, you know, some of the big ones (universities) the reason we are able to partner them is that we offer a non-Chinese qualification. It is not because they haven't got the materials there and all the equipment, and the expertise.

As Westerners we have got a very false view of what China is about, because we get to Beijing, we go to Shanghai, we see fantastic buildings, we see the broadband in the hotel room. All of that, but the vast majority of people are actually still living in villages with a mud floor, no running water, no electricity, no computers. But obviously, once people get access to computers, then through the Internet and everything, there is massive scope for the transfer of knowledge.

We can learn something and they can learn something, which is more the partnership model if you like. But certainly, some of them, with the equipment they have got will just blow us away.

In many places, for instance broadband access is certainly not that great. And students don't on the whole have computers at home like we have them at home. Even university staff, the senior staff would but I'm sure that the lower level staff would not necessarily have them at home. And they sometimes have them but their access to broadband might be quite weak.

Blocking and filtering problems were highlighted by 25% of the participants. Some institutions reported more problems than others and it seemed to be related to the cultural and political sensitivity of their work.

One of our big problems is that somebody is blocking our website. We are not sure what it is related to and it's been going on for about a year and a half. So we have had multiple people going up to talk to the Public Security Bureau, talks to key communications authorities and they say it has nothing to do with them, but the site is still blocked.

The number of people doing Internet security in China is huge. They have devoted an enormous amount of time to it, so they could easily do it. And the university would be looking at it to make sure you weren't doing anything regarded as illegal.

Yes, there are those sorts of control. They control it a lot more than we do here. I mean over time, I think governments will begin to control the Internet more rather than less. And by that, I mean all governments. Largely because they are getting a handle on the technology that will allow them to do it. I've never seen anything a government could control that it didn't.

Sorting out issues of blocking, filtering and censorship necessitate the development of strong cultural and political understandings on both sides. Some issues would seem innocuous by members of one culture but would be offensive or sensitive to members of another culture. One participant related a story concerning some files thought to be of little importance, and without apparent sensitivities were returned to the sender on the CD, marked not suitable for international sending. After checking by both sides, the files were not sensitive, and they were transferred by FTP (file transfer protocol).

Concerns about respecting intellectual property rights were flagged by several participants as a challenge arising from knowledge transfer between Australia and China.

> It is not difficult for the Chinese who simply do their level best to get the information and use it and not be constrained terribly much by international laws or propriety. But it is like anywhere else, there will always be propriety secrets, trade secrets. You will know what you have to know. Nobody is going to lay it all out to be knocked off by anyone.

> The most you can hope for with IP in China is for a five year agreement—after that, forget it.

A common theme raised by participants was that ICT should facilitate a genuine partnership, as there was tremendous value in a two-way transfer of knowledge.

> Of course ICT will facilitate transfer, there is no problem about that, but we have got to stay away from the idea that it should be a one-way transfer because we know everything. We don't. I mean they have tremendous knowledge. We can be certain that we will learn something and they will learn something.

The content analysis has confirmed the perception by Australian academics that ICT is a critical enabling factor in knowledge transfer. They reported that the use of ICT to facilitate knowledge transfer was taken up enthusiastically by their Chinese counterparts. This could be partly explained by the emphasis placed on science and technology within China. Recently, Ian Watt, the Australian embassy's minister-counsellor responsible for education, stated that "the Chinese government strongly believes that excellence in science and technology translates directly into strength in the economic and political arenas." Furthermore over half the university places in China are devoted to science and/or technology students compared to 10 percent in Australia (Callick, 2008). However, making the best use of ICT was not without challenges that need to be addressed in the future.

Limitations and future directions

A limitation of this research project was the small sample size. This project aimed to gather interviews from 25 participants: 20 Australians and five Chinese. This target was not met due to time and logistical constraints. It proved particularly difficult to obtain interviews with potential Chinese participants. This might be overcome in the future by traveling to China and conducting face-to-face interviews once sufficient rapport (e.g., trust) has been established. However, sufficient scope was obtained from the key Australian participants for this research to show specific trends and emerging themes. These key Australian participants were all highly qualified to discuss knowledge transfer issues between China and Australia because they have all worked extensively with the Chinese and have acquired a deep appreciation of the factors that contribute to successful knowledge transfer. In the future it would be ideal to interview at least 20 Australian and 20 Chinese participants in order to adequately compare and contrast viewpoints. A further limitation was that participants from business and industry were sought for this project; however, only academic participants were obtained. Where a balance between participants from business and industry is not necessary, having this additional viewpoint may provide further insight into how best to accomplish knowledge transfer. Due to the small sample size, generalization of these findings to the wider population should be undertaken with caution.

Conclusion

In conclusion, China is undergoing tremendous change, and part of those changes includes building world class higher educational institutions and participating in international knowledge transfer between academics and researchers. Currently, it appears that knowledge transfer is occurring between Australia and China, but it is generally one-sided (Australia to China). It has been suggested by the literature, and from the findings of this research project that increased cultural understanding, and the building of trust based relationships, is a prerequisite for China to genuinely participate in reciprocal transfer of knowledge. These findings are expected to contribute to the growing discourse on international, intercultural and bi-directional understanding of knowledge transfer. The benefit of this research is to gain a more comprehensive understanding of knowledge transfer. This is expected to enhance academic and research excellence and relevance between Australia and China, which may also be applicable to improve knowledge transfer between Australia and other countries.

United States / China Knowledge Exchange

The relationship between the United States and China and the resultant exchange of knowledge, along with technology transfer, has the potential to improve the lives of more people than any other global partnership. The United States remains one of the world's most technologically advanced and wealthy nations, whereas China, despite the phenomenal growth of its national economy, contains millions of people living in poverty or in very disadvantaged economic conditions. The main drivers of knowledge transfer between the nations has been trade, direct foreign investment and more recently, wholly owned U.S. companies operating in China.

In the year 2000, U.S. exports to China reached $16.3 billion (US$) and by 2007 this had increased to $65.2 billion. In 2000 the imports from China stood at 100 billion and this has increased to $321 billion in 2007 (The China–U.S. Business Council). In 2007, the category topping imports and exports was electrical machinery and equipment, demonstrating the importance of technology to both countries. China is currently the world's leading nation in terms of incoming direct foreign investment with a total of $82.7 billion. The U.S. contribution of $2.6 billion means that the U.S. is China's 6th largest source of direct foreign investment (The China–U.S. Business Council). This represents a decline in the influence of the U.S. through direct foreign investment. Some of the current global investment, however, does not represent true investment in China but a way of channeling funds through China and then back to other countries as a means of reducing tax. In recent years, the British Virgin Islands and Mauritius have greatly increased cash flow through China.

The level of financial success experienced by U.S. companies in China has greatly influenced the motivation towards facilitating knowledge transfer through formal education and training. U.S. based ICT company Motorola remains one of the leading foreign investors in China and provides tertiary level training at the Motorola University. For example, through this institution, U.S. universities offer their MBA courses to Chinese managers in China in addition to degrees relating to technical knowledge. The Motorola corporate profile is an example of a stated intention towards global responsibility and claims that "we harness the power of our global business to benefit society. Through our products, services and operations, we work to create economic opportunities and growth in regions where we do business. We know there is only one earth, so we foster sustainable use of the earth's resources in our products and operations, and we strive to design environmentally conscious products. We also know there are many compelling needs in the world. Through charitable giving and by expanding access to technology and the accompanying socio-economic benefits of our products, we invest in the many communities where we operate" (http://www. motorola.com/mot/doc/7/7206_MotDoc.pdf).

U.S. based multinationals such as Motorola have fostered a great deal of technology transfer to China.

Despite an international reputation concerning great risk associated with ventures in China, the China-U.S. council reports high levels of success for U.S. companies "Contrary to perceptions, FIEs in China are demonstrating good financial results. According to the USCBC 2007 Member Priority Survey, 83% of respondents indicated that their China operations had posted a profit, with roughly two-thirds saying their rate of profitability in China was the same or better than their company's global profit margins" (U.S.–China Business Council, 2008, p. 3). Healthy profits do not necessarily lead to benefits in terms of equity, but the experience in China supports the view that companies are more likely to foster knowledge exchange under these conditions.

Is the United States losing its high tech advantage?

American motivation for maintaining knowledge exchange with China has been centred on access to the large local market and the profits associated with this while China has coveted access to high tech knowledge often involving Information Communication Technologies. Buckley, Clegg and Tan (2004, p. 31) argue that "the recent strategy of the Government of China has been to obtain foreign direct investment in order to obtain foreign technology and capital." Jiang (2005) also wrote about the lack of high tech knowledge in Chinese ICT firms and their desire to transfer knowledge from foreign sources. These findings were consistent with the earlier seminal work by Yan and Gray (1994), who concluded that "firms form joint ventures in China primarily to penetrate the local market and to pursue financial goals. To them market share and profitability are the important measures of venture performance. In contrast, for Chinese partners the overwhelming goal for cooperating with the West is to learn the more advanced Western technology." Since the U.S. led the world in ICT development after World War II with advances such as microcomputers, semiconductors, computer operating systems and other software, network hardware and software, processors and biotechnology, they held a great advantage and bargaining chip to leverage access to the expanding local market.

Recently, Tassy (2008) has questioned the likelihood of a continuing dominance of the U.S. as the leading technology based economy. Any reduction in maintaining ICT cutting edge development would have a potentially negative affect on knowledge transfer between the U.S. and China. Tassy (2008, p.561) claims that "most Americans think of their economy as the dominant technology leader, but trends clearly show that U.S. investments in R & D [research and development] and more broadly in 'innovation infrastructure' are increasingly inadequate to maintain that position." As evidence, Tassey points to the leading position of the U.S. in R & D spending (as a percentage of the GDP) in 1964 compared to currently holding 7th position. He points to the lack of U.S. dominance in nanotechnology development as being a stark contrast to previous watershed technological developments as a marker of diminished importance. Tassey maintains that globalisation has lead to a great deal of international technology transfer and that has shifted the dynamics of technology associated competition. In his view the U.S. is living on past glories and is guilty of complacency, which will ultimately lead to other economies becoming dominant with new ICT advances and innovations coming from elsewhere.

Barriers and enablers of U.S.—China knowledge transfer

According to Li-Hua (2004, p. 108) knowledge transfer consists of explicit and tacit knowledge and that "explicit knowledge is in general transferred through formal means such as conferences, meetings, seminars and training sessions while tacit knowledge is transferred through informal means, such as on the job training, telephonic communication, social occasions and chance meetings." All of these aspects are features of knowledge exchange between China and the U.S. but success is not consistent across all partnerships. The U.S.–China business council provides evidence of substantial success in terms of profitability, but some problems with collaborations have been documented. Muthusamy, White and Carr (2007, p. 53) recorded that alliances "have been poor, unstable, ineffective and poorly performing" and that examples exist of opportunistic use of the alliances to learn business or technological secrets. What then are the enablers and barriers to successful alliances leading to substantial knowledge transfer or exchange?

Perhaps the most important enabler or barrier to successful knowledge exchange is the degree of congruence of overall strategic goals of the partners. If the goals are not congruent then they should at least be complementary. Yan and Gray (1994) regarded most U.S./Chinese alliances to have strongly contrasting strategic objectives. They argued that the objectives did complement each other but also set the stage for a wrestle over control. Chinese control was perceived to be a long term objective by the Chinese partners. Debate exists as to whether knowledge transfer is more effective when one of the partners exercises strong control or whether shared control produces better results. Yan and Gray (1994) concluded that joint control led to superior performance, whereas Calantone and Zhao (2001) argued that strong U.S. control led to superior performance due to the unstable market in China and the U.S. partners' advantage in technological and managerial expertise. More research on the effects of control dynamics within U.S./China partnerships is needed. Control is also mediated by the bargaining power that each partner can exercise at any given time.

Another important factor is the readiness and capacity of the Chinese partner to effectively use new knowledge in their work context. If the learning is mediated by formal training on the job or in tertiary institutions, cost becomes an important factor. Wang, Tong and Koh (2004, p. 175) warned that the costs associated with knowledge transfer can be substantial and that "costs may be even higher in cross-border transfers where differences in cultural, political and socio-economic aspects compound the difficulty of transfer." They contend that U.S. partners are willing to do this when the partnerships are considered to be important in the development, production and marketing of lead products. Partnerships ideally allow the pooling of resources in cost effective ways. Developing positive relationships with different levels of government is also an important factor in successful collaborations. Calantone and Zhao (2001) emphasise the importance of effectively working with government authorities, particularly local government in China.

Gassmann and Han (2004) claim that cultural barriers (including language barriers) can be overcome by enlisting the help of 'global Chinese.' They point to the huge Chinese community living in the U.S. as a great asset to partnerships and knowledge transfer. Many of these people speak Chinese and were educated in the United States and have an understanding of both cultures. They caution that appointments should not be made on the assumption that these characteristics alone would be enough to ensure success, since other requirements such as having the appropriate technical expertise and negotiating and communication skills are also necessary. They also point out other complicating issues such as salary differentials between these people and local Chinese, high expectations and the ability or inability to handle cultural conflict as important mediating factors.

The development of trust between partners has emerged as perhaps the most important enabler of successful knowledge transfer between the U.S. and China. Where this has been a feature of partnerships, other problems have been more easily resolved. Muthusamy, White and Carr (2007, p. 67) concluded from their study that "relational social exchanges such as reciprocity, interorganizational trust and relative mutual influence between partners are positively related to perceived effectiveness of the alliance and commitment to continue the alliance." Knowledge transfer is a common and increasing area of research in the domain of business and management but is often ignored in equity discussions. The success of knowledge and technology transfer between the U.S. and China, however, has potential benefits that far outweigh governmental or philanthropic programs aimed at bridging the digital divide.

■ ■ ■

References

Ackerman, J. (2006). *Motivation for writing through blogs*. Master of Arts thesis submitted to Graduate College, Bowling Green State University, Bowling Green, OH, USA.

Adya, M., & Kaiser, K.M. (2005). Early determinants of women in the IT workforce: A model of girls' career choices. *Information Technology & People, 18*(3), 230–259.

Agosto, D.E. (2004). Girls and gaming: A summary of the research with implications for practice. Retrieved 27 July, 2005, from: http://girlstech.douglass.rutgers.edu/PDF/GirlsAndGamingOld. pdfAkshay

AAIDD (2007). AAIDD Fact Sheets – Definition of MR Q & A about intellectual disability. American Association on Intellectual and Development Disability.

Akshay, M., & Dhirubhai, A. (2005). ICT and rural societies: Opportunities for growth. *The International Information & Library Review, 37*, 345–351.

Akubue, A. I. (2002). Technology transfer: A third world perspective. *The Journal of Technology Studies*, Winter–Spring. Retrieved October 3, 2007, from http:scholar.lib.vt.edu/JOTS/Winter–Spring2004/pdfakubue.pdf

Alavi, M., & Leidner, D. (2001). Knowledge management and knowledge management systems: Conceptual and research issues. *MIS Quarterly Reviews, 25*(1), 107–136.

Aldrich, C., Gibson, D., & Prensky, M. (2007). *Games and simulations in online learning: Research and development frameworks*. Hershey, PA: Idea Group.

American Association of University Women (AAUW). (2000). *Tech-savvy: Educating girls in the new computer age*. Washington, DC: American Association of University Women Educational Foundation.

Anderson, J., & van Weert, T. (2002). *Information and communication technology in education*. Paris: UNESCO.

Anderson, N. (2000). Technology in schools: Beyond the doomsters and boosters. *Enterprising Schools, 1*(8), 6–12.

Anderson, N. (2001). *Inclusion: Can teachers and technology meet the challenge?* Unpublished PhD thesis, QUT, Brisbane.

Anderson, N. (2003). Making the link: Three successful university, industry and community collaborations around ICT development. In J. Renner & J. Cross (Eds.). *Higher education without borders, sustainable development in higher education*. Perth, WA: Edith Cowan University.

Anderson, N. (2005). Building digital capacities in remote communities within developing countries. In D. Kamalavijayan & H. Parekh (Eds). *International Conference on Information Management in a Knowledge Society*, Mumbai, India, February 21–25: Allied Publishers.

Anderson, N., Klein, M., & Lankshear, C. (2005). Redressing the gender imbalance in ICT professions: Toward state-level strategic approaches. *Australian Educational Computing, 20*(2), 3–10.

Anderson, N., Lankshear, C., Courtney, L., & Timms, C. (2006). Girls and ICT survey: Initial findings. *Curriculum Leadership, 4*(12). Retrieved March 5, 2007, from http://cmslive.curriculum.edu.au/leader/default.asp?issueID=10270&id=13812

Anderson, N., Lankshear, C., Timms, C., & Courtney, L. (2008). 'Because it's boring, irrelevant and I don't like computers': Why high school girls avoid professionally-oriented ICT subjects. *Computers and Education, 50,* 1304–1318.

Anderson, N., Timms, C., & Courtney, L. (2006). "If you want to advance in the ICT industry, you have to work harder than your male peers." Women in ICT industry survey: Preliminary findings. Paper presented at the 10th Annual Australian Women in Information Technology Conference (AusWIT), December 5 & 6, Adelaide, SA, Australia.

Anderson, N., Timms, C. & Courtney, L. (2007). Rural and metropolitan comparisons, "Twelve months later, we are still waiting…": Insights from the Girls and ICT Study. *Journal of Media, Technology and Lifelong Learning, 3,* (3) Norway. Retrieved September 10, 2008, from http://www.seminar.net/current-issue/and-twelve-months-later-we-are-still-waiting

Anderson, Ronald E. (1993). *Computers in American schools, 1992: An overview. From the IEA Computers in Education Study*. Minneapolis, MN: University of Minnesota.

Annan, K. (2003). *Address to the World Summit on information technology,* 10 December 2003. Retrieved March 9, 2007, from http://www.un.org/apps/sg/ sgstats.asp?nid=695

Armstrong, J. (2005). *Is IT unfriendly to women?* Retrieved June 29, 2005, from http://techrepublic.com.com/5102-10878-5753934.html

Argote, L., & Ingram, P. (2000). Knowledge transfer: A basis for competitive advantage in firms. *Organizational Behaviour and Human Decision Processes, 82*(1), 150–159.

Ashton, T. M. (2000). Assistive technology. *Journal of Special Education Technology, 15*(1), 57–58.

Ashton, T. M. (2003). A totally hands-free computer experience: The smart-nav. *Journal of Special Education Technology, 18*(3), 60.

Australian Bureau of Statistics (ABS). (2006). Education and work, Australia, May 2006, 6227.0. Retrieved March 18, 2007, from http://www.abs.gov.au/ AUSSTATS/abs@.nsf/mf/6227.0? OpenDocument

Bandura, A. (1997). *Self-efficacy: The exercise of control.* New York: Freeman.

Battro, A. (2007). *One child per laptop Wiki.* Retrieved June 1, 2008, from http://wiki.laptop.org/ wkik/Current_events

Baumard, P. (1999). *Tacit knowledge in organizations.* Thousand Oaks, CA: Sage.

Beck, C., & Kosnick, C. (2006). *Innovations in teacher education: A social constructivist approach.* New York: State University of New York Press.

Becker, H. (2000). Findings from the teaching, learning, and computing survey: Is Larry Cuban right? Paper presented at the School Technology Leadership Conference, Washington, DC, January.

Becker, H., & Sterling, C. (1987). Equity in school computer use: National data and neglected considerations. *Journal of Educational Computer Research, 3*(3), 289–309.

Bender, A. (2005). To compete, South Korean universities step up use of English. *The Chronicle of Higher Education, 52*(17), pNA.

Bhatt, C. (2001). Knowledge management in organization: Examining the interactions between technologies, techniques, and people. *Journal of Knowledge Management, 5*(1), 68–75.

Bigum, C. (2002). Design sensibilities, schools, and the new computing and communications technologies, in I. Snyder (Ed.), *Silicon literacies.* London: Routledge-Falmer.

Bigum, C. (2003). The knowledge producing school. Moving away from finding educational problems for which computers are solutions. Retrieved September 10, 2008, from http://www.deakin. edu.au/education/lit/kps/pubs/ comp_in_nz.rtf

Bigum, C., & Kenway, J. (1998). New information technologies and the ambiguous future of schooling: Some possible scenarios. In A. Hargreaves, M. Lieberman, M. Fullan & D. Hopkins (Eds.), *International handbook of educational change* (375–395). Hingham, MA: Kluwer Academic.

Bishop, J. (2006). Knowledge transfer and engagement forum. Keynote Address, Darling Harbour, Sydney, 16 June. Retrieved June 2, 2008, from http://www.dest.gov.au/ministers.media/ bishop/2006/06/b001160606.asp

Balandin, S. & Duchan, J. (2007). Communication: Access to inclusion. Journal of Intellectual and Developmental Disability, 32(4), 230-232.

Barcellini, F., Detienne, F., & Burkhardt, J. (2008). Participation in online interaction spaces: Design-use mediation in open source software community. International Journal of Industrial Ergonomics. Available as corrected proof at: http://www.sciencedirect.com/science?_ob=ArticleURL&_udi=B6V31-4V38XN5-3&_user=972264&_rdoc=1&_fmt=&_orig=search&_sort=d&view=c&_acct=C000049659&_version=1&_urlVersion=0&_userid=972264&md5=664e56251890ebc2bc2bc02a277369ad

Bitzer, J., & Kerekes, M. (2008). Does foreign direct investment transfer technology across borders? New Evidence. Economic Letters, 100, 355–358.

Blacker, F. (1995). Knowledge, work and organizations: An overview and interpretation. Organizational Studies, 16(6), 1021–1046.

Blackstone, S. W. (1996). Selecting, using and evaluating communication devices. In J.C. Galvin & M. J. Scherer (Eds.), Evaluating, selecting and using appropriate assistive technology. Gaithersburg, MD: Aspen Publishers.

Blood, R. (2002). The weblog handbook: Practical advice on creating and maintaining your blog. Cambridge, M.A., Perseus Publications.

Boeije, H. (2002). A purposeful approach to the constant comparative method in the analysis of qualitative interviews. Quality and Quantity, 36, 391–409.

BoingBoing. (2007). One Laptop Per Child machines for sale this Christmas: buy two, one goes to developing world. Retrieved October, 6th, 2008 from http://www.boingboing.net/2007/09/23/one-laptop-per-child-1.html

Bond, R., & Castagnera, E. (2006). Peer Supports and Inclusive Education: An underutilized resource'. Theory into Practice, 45(3), 224-229.

Boone, R., & Higgins, K. (2003). Reading, writing and publishing digital text. Remedial and Special Education, 24(3), 132–139.

Bonaccorsi, A. & Rossi, C. (2003). Why open source software can succeed. Research Policy, 32, 1243–1258.

Braddock, D., Rizzolo, M. C., Thompson, M., & Bell, R. (2004). Emerging technologies and cognitive disabilities. Journal of Special Education Technology, 19(4), 49–56.

Brannen, J. (Ed.). (1992). Mixing methods: Qualitative and quantitative research. Aldershot, England: Thomas Coran Research Institute, Institute of Education.

Bransford, J., & Vye, N. (1989). A perspective on cognitive research and its implications for instruction. In L. Resnick & L. E. Klopfer (Eds.), Toward the thinking curriculum: Current cognitive research, Alexandria, VA: ASCD. 173–205.

Broos, A. (2005). Gender and information and communication technologies (ICT) anxiety: Male self assurance and female hesitation. Cyberpsychology and Behavior, 8(1), 21–31.

Brophy, J. (2006). Graham Nuthall and social constructivist teaching: Research-based cautions and qualifications. Teacher and Teacher Education, 22, 529–537.

Brown, D., Battersby, S., & Shopland, N. (1995). Design and evaluation of a flexible training environment for use in a supportive employment setting. *International Journal of Disability and Human Development, 4*(3), 251-258.

Brown, D., Shopland, N., Battersby, S., Lewis, J., & Evett, L. (2007). *Proceedings of the European Conference on Games-Based Learning,* Glasgow, UK, 37–46.

Bruns, D. & Mogharreban, C. (2007). The gap between beliefs and practices: Early childhood practitioner's perceptions about inclusion. *Journal of Research in Childhood Education, 21*(3), 229–241.

Bryant, D. P., & Bryant, B. R. (2003). *Assistive technology: For people with disabilities.* Boston, MA: Pearson Education.

Bryman, A. (1992). *Charisma and leadership in organisation.* London: Sage.

Buckley, P. J., & Carter, M. J. (2000). Knowledge management in global technology markets: Applying theory to practice. *Long Range Planning, 33,* 55–71

Buckley, P., Clegg, J. & Tan, H. (2004). Knowledge transfer to China: Policy lessons from foreign affiliates. *Transnational Corporations, 13*(1), 31–71

Caffrey, A., & McCrindle, R. (2004). *Developing a multimodal web application.* Paper presented at the 5th International Conference on Disability, Virtual Reality & Associated Technology, Oxford.

Calantone, R., & Zhao, Y. (2001). Joint ventures in China: A comparative study of Japanese, Korean, and U.S. partners. *Journal of International Marketing, 9*(1), 1–23.

Callick, R. (2008). China looks uneasily on. *The Australian Newspaper,* Higher Education, June 25, 28.

Calsyn, R., Quicke, J., & Harris, S. (1980). Do improved communication skills lead to increased self-esteem? *Elementary School Guidance & Counselling,* 5, (1), 49–55.

Camp, T. (1997). The incredible shrinking pipeline. *Communications of the ACM, 40*(10), 103–110.

Campbell, D. (1975). Degrees of freedom and the case study. *Comparative Political Studies, 8,* 178–193.

Cantu, E., & Farines, J. (2006). International Federation for Information Processing, Volume 210, in *Education for the 21ˢᵗ Century-Impact of ICT and Digital Resources,* Editors: D. Kumar and J. Turner.

Carey, A., & Friedman, M., & Bryen, D. (2005). Use of electronic technologies by people with intellectual disabilities. *Mental Retardation, 43*(5), 322–333.

Carpenter, T., & Fennema, E. (1992). Cognitively guided instruction: Building on the knowledge of students and teachers. *International Journal of Educational Research, 17*(5), 457–470.

Coar, K. (2006). The open source definition. Retrieved from http://www.opensource.org/docs/definition.php in November 2008.

Carroll, T. (2000). If we didn't have the schools we have today, would we create the schools we have today? *Contemporary issues in technology and teacher education.* 1(1). Retrieved September 10, 2008, from http://www.citejournal.org/vol1/iss1/currentissues/general/article1.htm

Carter, E. Lane, K., Pierson, M., & Stang, K. (2008). *Journal of Special Education Online*, May, n.p.

Central Intelligence Agency (CIA). (2008). *The 2008 CIA world factbook 2008,* Central Intelligence Agency, United States, Institute: Science, Engineering and Technology. Retrieved June 12, 2008, from http://www.intute.ac.uk/ sciences/ worldguide/ countrycompare.html

Christie, M. (2004). Computer databases and Aboriginal knowledge. *International Journal of Learning in Social Contexts*, *1*, 4–12.

Cisco Systems. (2002). *Gender initiative.* Retrieved July 6, 2005 from http://gender.ciscolearning. org

Cobb, P., Yackel, E., & Wood, T. (1992). A constructivist alternative to the representational view of mind in mathematics. *Journal for Research in Mathematics Education.* 23, (1), 2–33.

Cohoon, J.M. (2003). *Must there be so few? Including women in CS.* Paper presented at the 25th International Conference on Software Engineering, May. Retrieved July 25, 2005, from http:// portal.acm.org/results.cfm

Commonwealth of Australia. (2005). *Backing Australia's ability: Building our future through science and innovation.* Retrieved June 4, 2008, from http://backingaus.innovation.gov.au/info_booklet/ on_commit.htm

Commonwealth of Australia. (2008). *Our government.* Retrieved June 1, 2008 from http://www. australia.gov.au/Our_Government

Connell, R. & White, V. (Eds). (1991). *Running twice as hard— The Disadvantaged Schools Program in Australia.* Geelong, VIC: Deakin University Press.

Cook, A. M., & Hussey, S. M. (2002). *Assistive technologies: Principles and practice.* St Louis, MI: Mosby.

Cooney, G., & Jahoda, A., Gumley, A., & Knot, F. (2006). Young people with intellectual disabilities attending mainstream and segregated schooling: Perceived stigma, social comparison and future aspirations. *Journal of Intellectual Disability Research, 50*(6), 432–444.

Corbitt, B., Peszynski, K., Inthanond, S., Hill, B., & Thanasankit, T. (2004). Cultural differences, information and code systems. *Journal of Global Information Management, 2,* 65.

Courtney, L., Lankshear, C., Timms, C., & Anderson, N. (in press). Insider perspectives on the ICT industry: Findings from the Australian Women in ICT Industry Survey. Policy Futures in Education. (accepted).

Courtney, L., Timms, C., & Anderson, N. (2006). "I would rather spend time with a person than a machine". Qualitative findings from the girls and ICT survey. In A. Ruth (Ed.). *Quality and impact of Qualitative Research*, 3rd annual QualIT conference proceedings), Brisbane, QLD: Institute for Integrated and Intelligent Systems, Griffith University, 51–57.

Craddock, G. (2006). The AT continuum in education: Novice to power user. *Disability and Rehabilitation: Assistive Technology, 1*(1–2), 17–27.

Creswell, J. W. (1994). *Research design: Qualitative & quanitative approaches.* Thousand Oaks, CA: Sage Publications.

Cuban, L. (2000). So much high-tech money invested, so little use and change in practice: How come? Paper presented at the School Technology Leadership Conference, Washington DC: January.

Cuny, J., & Aspray, W. (2002). *Recruitment and retention of women graduate students in computer science and engineering.* Results of a workshop organized by the Computing Research Association, San Francisco, CA, June 21–22, 2000. *SIGCSE Bulletin, 34* (2), 168–174.

Daiute, C. (1992). Multimedia composing: Extending the resources of kindergarten to writers across the grades. *Language Arts, 69,* 250–260.

Dautenhahn, K. (2000). *Design issues on interactive environments for children with autism.* Paper presented at the 3rd International Conference on Disability, Virtual Reality & Associated Technology, Alghero, Italy.

Davenport, T., & Prusak, L. (1998). *Working knowledge: How organizations manage what they know.* Boston, MA: Harvard Business School.

Davies, D. K., Stock, S., & Wehmeyer, M. L. (2001). Enhancing independent internet access for individuals with mental retardation through use of a specialized web browser: A pilot study. *Education and Training in Mental Retardation and Developmental Disabilities, 36*(1), 107–113.

Davies, P. (1989). Technology in special education in Australia. *Australian Educational Computing, 3*(2), 32–39.

Dempsey, I. & Conway, R. (2005). Educational accountability and students with a disability in Australia. *Australian Journal of Education, 49*(2), 152–168.

De Palma, P. (2001). Viewpoint: Why women avoid computer science. *The Communications of the ACM, 44*(6), 27–29.

De Palma, P. (2005). The software wars. *The American Scholar, 74*(1), 69–83

De Palma, P. (2006). Women, mathematics and computing. In E. Trauth (Ed.). *Encyclopaedia of gender and information technology,* Pittsburgh, PA: Idea Group Reference, 1303–1308.

Department of Communications, Information Technology and the Arts (DCITA). (2005). *Participation summit: Overview paper.* Retrieved March 12, 2007, from http://archive.dcita.gov.au/ 2005/09/ participation_summit/ breakout_groups/overview_paper

Department of Education, Queensland. (1993). *Educational provisions for students with disabilities: Policy statement and management plan.* Brisbane, QLD.

Department of Education and the Arts, Tasmania. (1994). *Inclusion of students with disabilities in regular schools.* Hobart, TAS.

Department of Education, Science and Training (DEST). (2006). *Knowledge transfer and Australian universities and publicly funded research agencies* (Vol. 1), Phillips KPA Pty Ltd.

Department of Education Science and Training (DEST). (2006). *Students 2005 [full year]: Selected higher education statistics: Summary analyses.* Retrieved March 18, 2007, from http://www.dest. gov.au

Dertouzous, M. (2001). *The unfinished revolution: Human centered computers and what they can do for us.* New York: Harper Collins.

Disability Rights Commission. (2004). *The Web: Access and inclusion for disabled people — A formal investigation conducted by the Disability Rights Commission.* London: Disability Rights Commission.

Downing, J. E. (2000). Augmentative communication devices: A critical aspect of assistive technologies. *Journal of Special Education Technology, 15*(3), 35–38.

Dravis, P. (2003). *Open source software: Perspectives for development.* Washington DC: InfoDev, The Dravis Group. Retrieved June 6, 2006, from http://www.infodev.org

Drori, G., & Jang, Y. (2003). The global digital divide. *Social Science Computer Review, 2*(2), 144–161.

Duffy, P., & Bruns, A. (2006). The use of blogs, wikis and RSS in education: A conversation of possibilities. *Proceedings of Online Learning and Teaching Conference*, Brisbane.

Education Queensland. (2004). *Assistive technologies.* Retrieved July 20, 2006, from http://www.learningplace.com.au/deliver.content.asp?pid=16627

Education Queensland (EQ). (2006). Getting the best OP. *Schools + Parents, 1,* 13.

Edyburn, D. L. (2004). Rethinking assistive technology. *Special Education Technology Practice, 5*(4), 16–23.

Egbu, C. (2000). Knowledge management in construction SMEs: Coping with the issues of structure, culture, commitment and motivation. *Proceedings of the 16ᵗʰ Annual Conference of Association of Researchers in Construction Management (ARCOM)*, September 6 – 8, Glasgow Caledonian University.

Elerton, N. & Clements, M. (1992). Some pluses and minuses of radical constructivism in mathematics education. *Mathematics Education Research Journal, 4*(2), 3–19.

el Kaliouby, R., & Robinson, P. (2005). The emotional hearing aid: an assistive tool for children with Asperger syndrome. *Universal Access in the Information Society, 4*(2), 121–134.

Elliot, L. B., Foster, S., & Stinson, M. (2003). A qualitative study of teacher's acceptance of a speech-to-text transcription system in high school and college classrooms. *Journal of Special Education Technology, 18*(3), 45–59.

Eloff, I. & Kgwete, L. (2007). South African teachers' voices on support in inclusive education. International Focus Issue, *Childhood Education,* 351–356.

Epstein, J. N., Willis, M. G., Conners, C. K., & Johnson, D. E. (2001). Use of a technological prompting device to aid a student with Attention Deficit Hyperactivity Disorder to initiate and complete daily tasks: An exploratory study. *Journal of Special Education Technology, 16*(1), 19.

Ernst, D. & Lundvall, B. (2004). Information technology in the learning economy: Challenges for developing countries. In E. Reinert (Ed), *Globalization, economic development and inequality.* Northampton, MA: Edward Elgar Publishing.

e-skills UK. (2006). Research notes. Retrieved January 18, 2007, from http://www.e-skills.com

Etcheidt, S. (2006). Least restrictive and natural environments for young children with disabilities. *Topics in Early Childhood Special Education, 26*(3), 167–177.

European Commission. (2007). *Improving knowledge transfer between research institutions and industry across Europe.* Belgium: European Commission. Retrieved, May 22, 2008, from http://ec.europa. eu/invest-in-research/pdf/ download_en/knowledge_transfe_07. pdf

European Commission Information Society Directorate General. (2000). *Study on technology trends and future perspectives within assistive technologies.* Retrieved July 20, 2006, from http://www. cordis.lu/ist/ka1/ special_ needs/library.htm

Fallah, H., & Ibrahim, S. (2004). Knowledge spillover and innovation in technological clusters. *Proceeding of the 2004 International Association for Management of Technology* (pp. 1–16). Washington, DC. Retrieved June 6, 2008, from http://personal.stevens.edu/~hfallah/ KNOWLEDGE_SPILLOVER_AND_INNOVATION_IN_TECHNOLOGICAL_ CLUSTERS.pdf

Fantuzzo, J., & King, A.,& Heller, L. (1992). Effects of reciprocal peer tutoring on mathematics and school adjustment: A component analysis. *Journal of Educational Psychology, 84*(3), 331–339.

Felsenstein, L. (2005). Problems with the OPCL approach. Retrieved from: http://fonly.typepad. com/fonlyblog/2005/11/problems_with_t.html Oct, 6th, 2008.

Ferguson, D. (2008). International trends in inclusive education: the continuing challenge to teach each one and everyone. *European Journal of Special Needs Education, 23*(2), 109–120

Filler, J., & Xu, Y. (2007). Including children with disabilities in early childhood education programs. *Childhood Education,* Winter, 92–98.

Flor, A. (2001). ICT and poverty: The indisputable link. *Proceedings of The Third Asia Development Forum on Regional Economic Cooperation in Asia and the Pacific.* June 11–12, Bangkok, Thailand.

Foreman, J., & Borkman, T. (2007). Learning sociology in a massively multistudent online environment. In Gibson, D, Aldrich, C., & Prensky, M. (Eds) *Games and simulations in online learning: Research and development frameworks.* Hershey PA. London.

Forlin, C., & Keen, M., & Barrett, E. (2008). The concerns of mainstream teachers: Coping with inclusivity in an Australian context. *International Journal of Disability, Development and Education, 55*(3), 251–264.

Frenkel, K.A. (1990). Women and computing. *Communications of the ACM Journal, 33 (11),* reprinted at website of Computer Professionals for Social Responsibility (CPSR). Retrieved June 29, 2005, from http://www.cpsr.org/issues/womenintech/frenkel-cacm-womcomp

Frieden, R. (2005). Lessons from broadband development in Canada, Japan, Korea and the United States. *Telecommunications Policy, 29*, 595–613.

Frieze, C. (2005). *Diversifying the images of computer science: Undergraduate women take on the challenge!* Paper presented at the Special Interest Group of Computer Science Education '05 (SIGCSE), February 23–27, St Louis, MO. Retrieved June 29, 2005, from http://delivery.acm. org

Fuchs, C., & Horak, E. (2008). Africa and the digital divide. *Telematics and Infomatics 25*, 99–116.

Gall, M., Borg, W. & Gall, J. (1966). *Educational research: An introduction* (6th ed.). White Plains, NY: Longman.

Gardner, J., & Bates, P. (1991). Attitudes and attributions on use of microcomputers in school by students who are mentally handicapped. *Education and Training in Mental Retardation*, March, 98–107.

Gassman, O., & Han, Z. (2004). Motivations and barriers of foreign R&D activities in China. *R&D Management*, *34*(4), 423–437.

Gates, J.L. (2002). *Women's career influences in traditional and nontraditional fields*. Poster presented at the Biennial meeting of the Society of Research in Adolescence, New Orleans, April 11–14). Retrieved January 12, 2006 from Eric database CE083 884.

Gee, J. (2003). What video games have to teach us about learning and literacy. New York: Palgrave Macmillan.

Gee, J. (2005). *An introduction to discourse analysis*. London: Routledge.

Gee, J.P. (2007). *Good video games and good learning: Collected essays on video games, learning and literacy (New literacies and digital epistemologies)*. New York: Peter Lang Publishing.

Gilbert, M.R., Masucci, M., Homko, C., & Bove, A.A. (2008). Theorizing the digital divide: Information and communication technology use frameworks among poor women using a telemedicine system. *Geoforum, 39*, 912–925.

Girls into doing great Information Technology Society (GIDGETS). (2007). *Go go GIDGETS*. Retrieved April 9, 2007, from http://www.learningplace.com.au/defaulteqa2.asp?orgid=48& suborgid=535

Goffee, R., & Jones, G. (2000). Why should anyone be led by you? *Harvard Business Review*, September–October, 62–70.

Goldfarb, A., & Prince, J. (2008). Internet adoption and usage patterns are different: implications for the digital divide. *Information Economics and Policy, 20*, 2–15.

Gorriz, C.M., & Medina, C. (2000). Engaging girls with computers through software games. *Communications of the ACM, 43*(1), 42–49.

Gourlay, S. (2002). Tacit knowledge, tacit knowing or behaving? *Proceedings of the Third European Conference on Organizational Knowledge, Learning and Capabilities*, April. Athens, Greece. Retrieved June 1, 2008 from http://www.alba.edu.gr/OKLAC2002/Proceedings/pdf_files/ ID269.pdf

Grace, R., Llewellyn, G., Wedgwood, N., Fenech, M., & Connell, D. (2008). Far from ideal: Everyday experience of mothers and early childhood professionals negotiating an inclusive early childhood experience in the Australian context. *Topics in Early Childhood Special Education, 5*, 18–31.

Granger, C.A., Morbey, M.L., Lotherington, H., Owston, R.D., & Wideman, H.H. (2002). Factors contributing to teachers' successful implementation of IT. *Journal of Computer Assisted Learning, 18*, 480–488.

Grant, R. M. (1996). Toward a knowledge based theory of the firm. *Strategic Management,*17 (special issue), 109–122.

Grassman, L. (2002). Identity and augmentative and alternative communication. *Journal of Special Education Technology, 17*(4), 41–44.

Green, J. C., & Caracelli, V. J., & Graham W. F. (1989). Towards a conceptual framework for mixed-method evaluation designs. *Educational Evaluation and Policy Analysis, 11*(3), 255–274.

Grey, J. (2004). Technology and social inclusion: Rethinking the digital divide by Mark Warschauer, *Journal of Economic Issues, 38*(1), 294–296.

Gurer, D., & Camp, T. (2002). An ACM-W literature review on women in computing. *Inroads SIGCSE Bulletin, 34*(2), 121–127.

Guynup, S., & Demmers, J. (2005). Fake fun: Transforming the challenges of learning to play. *International Conference on Computer Graphics and Interactive Techniques*, Los Angeles, CA.

Hartnell-Young, E. (2006). Teachers' roles and professional learning in community of practice supported by technology in schools. *Journal of Technology and Teacher Education, 14*(3), 461–480.

Haruvy, E., Prasad, A., & Sethi, S. (2003). Harvesting altruism in open-source software development. *Journal of Optimization Theory and Applications, 118*(2), 381–416.

Hasselbring, T. S., & Williams-Glaser, C. H. (2000). Use of computer technology to help students with special needs. *The Future of Children: Children and Computer Technology, 10*(2), 102–122.

Hastings, R., Sonuga-Barke, E., & Remington, B. (1993). An analysis of labels for people with learning disabilities. *British Journal of Clinical Psychology, 32*, 463–465.

Healy, A., & Connolly, T. (2007). Does games-based learning, based on constructivist pedagogy, enhance the learning experience and outcomes for the student compared to traditional didactic pedagogy? *Proceedings of the European Conference on Games-Based Learning*, Glasgow, 105–114.

Helmstetter, E., Peck, C., & Giangrego, M. (1994). Outcomes of intereactions with peers with moderate to severe disabilities: A statewide survey of high school students. *Journal of the Association of Persons with Severe Handicaps, 19*, 263–276.

Hermans, R., Tondeur, J., van Braack, J. & Valcke, M. (in press). The impact of primary school teachers' educational beliefs on the classroom use of computers. *Computers & Education*.

Hetman, F. (1973). *Society and the assessment of technology*. OECD: Paris.

Higgins, E. L., & Raskind, M. H. (2000). Speaking to read: The effects of continuous vs discrete speech recognition systems on the reading and spelling of children with learning disabilities. *Journal of Special Education Technology, 15*(1), 19

Higgins, E. L., & Raskind, M. H. (2005). The compensatory effectiveness of the Quicktionary Reading Pen II on the reading comprehension of students with learning disabilities. *Journal of Special Education Technology, 20*(1), 29–38.

Hill, B. (2007). *One child per laptop Wiki*. Retrieved June 1, 2008, from http://wiki.laptop.org/wkik/Current_events

Hines, R. A., & Hall, K. S. (2000). Assistive technology. *Journal of Special Education Technology, 15*(4), 37–39.

Hitchcock, C. (2001). Balanced instructional support and challenge in universally designed learning environments. *Journal of Special Education Technology, 16*(4), 23–30.

Hitchcock, C., & Stahl, S. (2003). Assistive technology, universal design, universal design for learning: Improved learning opportunities. *Journal of Special Education Technology, 18*(4), 45–52.

Hitt, L., & Tambe, P. (2007). Broadband adoption and content consumption. *Information Economics and Policy, 19*, 362–378.

Hocking, C. (1999). Function or feelings: factors in abandonment of assistive devices. *Technology and Disability, 11*(1–2), 3–11.

Hodas, S. (1996). *Technology refusal and the organizational culture of schools* (2nd ed.). Orlando, FL: Academic Press.

The Horizon Report. (2008). *The new media consortium.* Retrieved March 1, 2008, from http://www.nmc.org/horizon

Howells, J. R. (1995). Going global: The use of ICT networks in research and development. *Research Policy, 24*, 169–184.

HREOC. (2002). World Wide Web access: Disability Discrimination Act advisory notes: Australian Human Rights and Equal Opportunity Commission.

Huette, S. (2006). Blogs in education. Teaching Effectiveness Program: Be free to teach. Retrieved March 1, 2008, from http://eff.org/bloggers

ICT in Schools: The impact of government initiatives five years on. (2004). London: Ofsted Publications Centre.

Integrating ICTs into education: Lessons learned. A collective case study of six Asian countries. (2004). Asia and Pacific Regional Bureau for Education, UNESCO.

Interoperability Clearinghouse Glossary of Terms. (2008). *Knowledge.* Retrieved June 1, 2008, from http://www.ichnet.org/glossary.htm

Jackson, V. L. (2003). Technology and Special Education: Bridging the Most Recent Digital Divide: ERIC Document Reproduction Service No. ED 479685.

Jaffe, A. (1989). Real effect of academic research. *American Economic Review, 79*, 957–970.

Jaffe, A. B., Trajtenberg, M., & Fogarty, M. R. (2000). Knowledge spillovers and patent citations: Evidence from a survey of inventors. *American Economic Review, 9*(2), 215–219.

James Cook University, ARACY Submission, (2007). Unpublished.

Jeffs, T., & Morrison, W. F. (2005). Special education technology addressing diversity: A synthesis of the literature. *Journal of Special Education Technology, 20*(4), 19–25.

Jepson, A., & Perl, T. (2002). Priming the pipeline. *Inroads SIGCSE Bulletin, 34*(2), 36–39.

The Jhai Foundation (2006). Retrieved June 4, 2006, from http://www.jhai.org

Jiang, C. (2005). The impact of entrepeneur's social capital on knowledge transfer in Chinese high tech firms: the mediating effects of absorptive capacity and guanxi development. *Journal of Entrepreneurship and Innovation Management, 5*(3–4), 269–283.

Johnson, D., & Johnson, R. (1986). Computer assisted cooperative learning. *Educational Technology, 26*(1), 12–18.

Johnson, I. (2005). The global challenge: Strategies for the World Bank group. From predicting the present. *Harvard International Review, 27*(3), Fall. Retrieved June 27, 2006, http://hir.harvard.edu/articles/print.php?article=142

Johnson, J. (2003). Children, robotics and education. *Artif Life Robotics, 7*, 16–21.

Johnstone, B. (2002). *Discourse analysis.* Oxford: Blackwell Publishing.

Johnstone, S. S. (2003). Making the most of single switch technology: A primer. *Journal of Special Education Technology, 18*(2), 47–50.

Jonassen, D. (1996). *Computers as mindtools for schools: Engaging critical thinking.* Merrill, Upper Saddle River, N.J.

Jonassen, D. (2000). *Computers as mindtools for schools: Engaging critical thinking.* Columbus, OH: Prentice-Hall.

Jorgensen, C., & McSheehan, M., & Sonnenmeier, R. (2007). Presumed competence reflected in the educational programs of students with IDD before and after the Beyond Access professional development intervention. *Journal of Intellectual & Developmental Disability, 32*(4), 248–262.

Joseph, L., & Konrad, M. (2008). Teaching students with intellectual or developmental disabilities to write: A review of the literature. *Research in Developmental Disabilities.*

Kalyanpur, M., & Kirmani, M. H. (2005). Diversity and technology: Classroom implications of the digital divide. *Journal of Special Education Technology, 20*(4), 9–18.

Kardan, K. (2006). Computer role-playing games as a vehicle for teaching history, culture and language. *Sandbox Symposium Proceedings.* July, Boston, MA.

Kayess, R. & French, P. (2008). Out of the darkness into the light? Introducing the convention on the rights of persons with disabilities. *Human Rights Law Review, 8*(1), 1–34.

Khaled, R., Barr, P., Fischer, R., & Noble, J. (2006). *OZCHI 2006 Conference Proceedings.* November, Sydney, NSW, Australia.

King, T. W. (1999). *Assistive technology: Essential human factors.* Boston, MA: Allyn and Bacon.

Knight, S. (2003). Understanding the nature of knowledge for building effective knowledge management systems: Bridging the gap between cyber and cognitive space. *Proceedings of the 4th International Er-B Conference*, 24–25 November, Perth, Australia. Retrieved August 24, 2007 from http://www-business.ecu.edu.au/schools/mis/ media/pdf/0087.pdf

Kroker, A,, & Weinstein, M. (1994). *Data trash.* New York: St Martin's Press.

Kulik, C., Kulik, J., & Bangert - Drowns, R. (1985). Effectiveness of computer based instruction in elementary schools. *Computers in Human Behavior, 1*, 59–74.

Kuroda, T., Tabata, Y., Goto, A., Ikuta, H., & Murakami, M. (2004). *Consumer price data-glove for sign language recognition.* Paper presented at the 5th International Conference on Disability, Virtual Reality & Associated Technology, Oxford.

Lahm, E. A., & Sizemore, L. (2002). Factors that influence assistive technology decision making. *Journal of Special Education Technology, 17*(1), 15–26.

Lam, A. (2000). Tacit knowledge, organizational learning and societal institutions: An integrated framework. *Organizational Studies, 21*(3), 487–513.

Lane, J. (2005). The digital divide: Are our girls falling through the gap? *Australian Educational Computing, 20*(2), 11–15.

Lankshear, C. (1997). *Changing literacies.* Buckingham & Philadelphia, PA: Open University Press.

Lankshear, C., & Knobel, M. (2006). Blogging as participation: The active sociality of a new literacy. Paper presented to the *American Educational Research Association*, San Francisco, CA.

Lankshear, C., Synder, I., & Green, B. (2000). Teachers and techno-literacy. Managing literacy, technology and learning in schools. St. Leonards, NSW: Allen and Unwin.

La Rose, R., Gregg, J.L., Strover, S., Straubhaar, J., & Carpenter, S. (2007). Closing the rural broadband gap: Promoting adoption of the Internet in rural America. *Telecommunications Policy, 31,* 359–373.

Lathan, C. E., & Malley, S. (2001). *Development of a new robotic interface for telerehabilitation.* Paper presented at the 2001 EC/NSF Workshop on Universal Accessibility of Ubiquitous Computing, Alcacer do Sal, Portugal.

Lau, K., Erwin, B., & Petrovic, P. (1999). Creative learning in school with LEGO. *Proceedings of the Frontiers in Education Conference.* Puerto Rico.

Lee, H. & Cho, Y. (2007). Factors affecting problem finding depending on degree of structure of problem situation. *The Journal of Educational Research,* Nov–Dec, 113–120.

Lee, Y., & Vega, L. A. (2005). Perceived knowledge, attitudes and challenges of AT use in special education. *Journal of Special Education Technology, 20*(2), 60–63.

Leonard-Barton, D. (1995). *Wellspring of knowledge: Building and sustaining the resources of innovation.* Boston, MA: Harvard University Press.

Lepper, M., & Gurner, J. (1989). Children and computers: Approaching the twenty-first century. *American Psychologist, 2,* 170–178.

Leung, P., Owens, J., Lamb, G., Smith, K., Shaw, J., & Hauff, R. (1999). *Assistive technology: Meeting the technology needs of students with disabilities in post-secondary education.* Canberra, ACT: Commonwealth of Australia.

Leventhal, J. D. (1996). Assistive devices for people who are blind or have visual impairments. In J. C. Galvin & M. J. Scherer (Eds.), *Evaluating, selecting and using appropriate assistive technology.* Gaithersburg, MD: Aspen Publishers.

Li, E., Tam, A., & Wongman, D. (2006). Exploring the self-concepts of persons with intellectual disabilities. *Journal of Intellectual Disabilities, 10*(1), 19–34.

Li, P., Faitam, S., & Man, K. (2006). Exploring the self-concepts of persons with intellectual disabilities. *Journal of Intellectual Disabilities, 10*(1), 19-34.

Li-Hua, R. (2003). From technology transfer to knowledge transfer – A study of international joint venture projects in China. *Proceedings of the International Association of Management Technology (IAMOT) Conference,* Nancy, France. Retrieved August 20, 2007, from http://www.iamot.org/paperarchive/li-hua.pdf

Li-Hua, R. (2004). *Technology and knowledge transfer in China.* Ashgate Publishing, Aldershot, UK.

Li-Hua, R. (2007). Knowledge transfer in international educational collaboration programme: The China perspective. *Journal of Technology Management in China, 2*(1), 84–97.

The LINCOS Project (2006) Retrieved June4, 2006, from http://www.lincos.net/ webpages/ english/ general.html

Lindh, J., & Holgerson. (2007). Does Lego training stimulate pupils' ability to solve logical problems? *Computers & Education, 49*, 1097–1111.

Li-Tsang, C., Lee, M., Yeung, S., Siu, A., & Lam, C. (2006). A 6-month follow-up of the effects of an information and communication technology (ICT) training programme on people with intellectual disabilities. *Research in Developmental Disabilities, 28*, 559–566.

Li-Tsang, C., & Yeung, S., & Chan, C., & Hui-Chan, C. (2005). Factors affecting people with intellectual disabilities in learning to use computer technology. *International Journal of Rehabilitation Research, 28*(2), 127–133.

Lunn, J., Lalic, M., Smith, B., & Taylor, C. (2006). *A political and economic introduction to China. House of Commons Library*, Research Paper 06/36. Retrieved March 12, 2008 from http://www. parliament.uk/commons/ lib/research/rp2006/rp06-036.pdf

Lynn, K.M., Raphael, C., Olefsky, K., & Bachen, C.M. (2003). Bridging the gender gap in computing: An integrative approach to content design for girls. *Journal of Educational Computing Research, 28*(2), 143–162.

MacArthur, C. (1996). Using technology to enhance the writing processes of students with learning disabilities. *Journal of Learning Disabilities, 29*, 344–35.

Maglitta, J. (1995). Smarten up! *Computerworld, 2*(23), 84–86.

Male, M. (2003). *Technology for inclusion.* Boston, MA: Pearson Education.

Malone, T., & Lepper, M. (1987). Making learning fun: A taxonomic model of intrinsic motivations for learning. In R. Snow & M. Farr (Eds.), *Aptitude, learning, and instruction, Volume 3: Cognitive and Affective Process Analysis*, Hillsdale, NJ: Erlbraum.

Malumud, O., & Pop-Eleches, C. (2008). The effect of computer use on child outcomes. *Harris School Working Paper Series* 08.12, University of Chicago.

Manetti, M., Schneider, B., & Siperstein, G. (2001). Social acceptance of children with mental retardation: Testing the contact hypothesis with an Italian sample. *Journal of Behavior Development, 25*, 279–286.

Marchand-Martella, N., & Martella, R. (1993). Evaluating the instructional behaviors of peers with mild disabilities who served as first-aid instructors for students with moderate disabilities. *Child & Family Behavior Therapy, 15*(4), 1–17.

Margolis, J., & Fisher, A. (2003). *Unlocking the clubhouse: Women in computing.* Cambridge, MA: Massachusetts Institute of Technology Press.

Marks, P. (2006). Gadgets get the feel of the tactile world. *New Scientist, 1*(2560), 26.

Marsh, H. (1990). Self-description questionnaire Manual, University of Western Sydney, Macarthur.

Martin, D., Martin, M., & Carvalho, K. (2008). Reading and learning-disabled children: Understanding the problem. *The Clearing House*, 81(3), 113–121.

Massof, R. W. (2003). *Auditory assistive devices for the blind.* Paper presented at the 2003 International Conference on Auditory Display, Boston, MA.

Mathur, A., & Ambani, D. (2005). ICT and rural societies: Opportunities for growth. *The international information and Library Review, 37,* 245–351.

Matson, E., & DeLoach, S. (2004). Using robots to increase interest of technical disciplines in rural and underserved schools. *Proceedings of the 2004 American Society for Engineering Education Annual Conference and Exposition,* Kansas City, KS.

McAdam, R., Mason, B., & McCrory, J. (2007). Exploring the dichotomies within the tacit knowledge literature: Towards a process of tacit knowing in organizations. *Journal of Knowledge Management, 11*(2), 43–59.

McCulloch, P., & MacMahon, T. (1998). *What is?* Retrieved July 10, 2006, from http://education. qld.gov.au/curriculum/learning/students/disabilities/resources/information/at/atsn-46.html

McFayden, G. M. (1996). Aids for hearing impairment and deafness. In J. C. Galvin & M. J. Scherer (Eds.), *Evaluating, selecting and using appropriate assistive technology.* Gaithersburg, MD: Aspen Publishers.

McInerney, C. (2002). Knowledge management and the dynamic nature of knowledge. *Journal of the American Society for Information Science and Technology, 53* (12), 1009–1018.

McLaren, J. & Zappala, G. (2002). The new economy revisited: An initial analysis of the digital divide among financially disadvantaged families. Camperdown, NSW: The Smith Family.

McNair, S., Kirova-Petrova, A., & Bhargava, A. (2001). Computers and young children in the classroom: Strategies for minimizing gender bias. *Early Education Journal, 29*(1), 51–55.

McTavish, M. (2008). "What were you thinking?" The use of metacognitive strategy during engagement with reading narrative and informational genres. *Canadian Journal of Education, 31*(2), 405–430.

Means, B., & Knapp, M. (1991). Cognitive approaches to teaching advanced skills to educationally disadvantaged students. *Phi Delta Kappa, 12,* 282–288.

Meelissen, M., & Drent, M. (2007). Gender differences in computer attitudes: Does school matter? *Computers in Human Behaviour, 24,* 969–985.

Merbler, J. B., Hadadian, A., & Ulman, J. (1999). Using assistive technology in the inclusive classroom. *Preventing School Failure, 43*(3), 113–117.

Michaels, C. A., & McDermott, J. (2003). Assistive technology integration in special education teacher preparation: Program coordinators' perceptions of current attainment and importance. *Journal of Special Education Technology, 18*(3), 29–37.

Millar, N. (2006, November 14). IT women in the industry. *The Sydney Morning Herald.* Retrieved March 5, 2006, from http://www.smh.com.au/news/technology/it-women-in-the-industry/20 06/11/13/1163266481822.html

Millar, J., & Jagger, N. (2001). *Women in ITEC courses and careers.* London, UK Department of Education and Skills, Department for Employment. The Women's Unit: 156.

Miller, D. P. (1998). Assistive robotics: An overview. In V. O. Mittal (Ed.), *Assistive technology and AI,* Berlin: Springer-Verlag, 126–136.

Ministerial Council on Education, Employment, Training and Youth Affairs (MCEETYA). (2003). *Demand and supply of primary and secondary school teachers in Australia, part F, complementary research*. Retrieved September 1, 2006, from http://www.mceetya.edu.au/verve/_resources/part_f.pdf

Moore, J. A., & Teagle, H. F. B. (2002). An introduction to cochlear implant technology, activation and programming. *Language, Speech and Hearing Services in Schools, 33*, 153–161.

Moore, K., Griffiths, M., & Richardson, H. (2005). *Moving in, moving up, moving out? A survey of women in ICT*. Paper presented at Symposium on gender and ICT: Working for Change. Retrieved July 27, 2005 from: http://www.isi.salford.ac.uk/gris/wini/Publications/Symposium_on_Gender_and ICT_Main_paper_pdf.pdf

Moran, S. (2004, 26 November). Women mentors put student lawyers in the picture. *Sydney Morning Herald*. Retrieved March 18, 2007, from http://www.law.unsw.edu.au/news_and_events/News.asp

Moreno, J. & Saldana, D. (2005). Use of a computer-assisted program to improve metacognition in persons with severe intellectual disabilities. *Research in Developmental Disabilities, 26*, 341–357.

Muthusamy, S., White, M. & Carr, A. (2007). An empirical examination of the role of social exchanges in alliance performance. *Journal of Managerial Issues*, XIX, 1, 53–75

Myers, C. (2007). "Please listen, it's my turn": Instructional approaches, curricula and contexts for supporting communication and increasing access to inclusion. *Journal of Intellectual and Developmental Disability, 32*(4), 263–278.

Negroponte, N. (2006). *60 Minutes* transcript. Retrieved June 1, 2008, from http://www.olpctalks.com/nicholas_negroponte/olpc_60_minutes_interview.html

Newell, A. F. (2003). Inclusive design or assistive technology. In J. Clackson, R. Coleman, S. Keates & C. Lebbon (Eds.), *Inclusive design: Design for the whole population,* London: Springer-Verlag, 172–181.

Newmarch, E., Taylor-Steele, S., & Cumpston, A. (2000). *Women in IT—What are the barriers?* Conference paper presented to the Network of Women in Further Education Conference 'Net Gains: Women, Information Technology and Emerging Issues', Retrieved March 22, 2006, from www.dest.gov.au/ research/pubs/womeninit.htm

Noble, D. (1999) *The religion of technology: The divinity of man and the spirit of religion*. Harmondsworth, Middlesex, England: Penguin Books.

Nonaka, I. (1991). The knowledge-creating company. *Harvard Business Review* (November – December), 90–104.

Nonaka, I. (1994). A dynamic theory or organizational knowledge creation. *Organizational Science, 5*(1), 14–37.

Nonaka, I., & Takeuchi, H. (1995). *The knowledge-creating company*. Oxford: Oxford University Press.

Nowicki, E., & Sandieson (2002). A meta-analysis of school-age childrens' attitudes towards persons with physical disabilities or intellectual disabilities. *International Journal of Disability, Development and Education, 49*(3), 243-265.

O'Dell, C., & Grayson, C., Jr. (1998). *If only we knew what we know.* New York: The Free Press.

OFSTED (2004). ICT in schools: the impact of government initiatives five years on. London: Office for Standards in Education.

Ono, H., & Zavodny, M. (2007). Digital inequality: A five country comparison using microdata. *Social Science Research, 36,* 1135–1155.

Organisation for Economic Co-operation and Development (OECD). (2001). *Understanding the digital divide,* Paris: Author.

Organization for Economic Cooperation and Development (OECD). (2008). Tertiary education for the knowledge society: OECD thematic review of tertiary education: Synthesis report. OECD.

O'Rourke, J., & Houghton, S. (2006). Students with mild disabilities in regular classrooms: The development and utility of the student perception of classroom support scale. *Journal of Intellectual and Developmental Disability, 31*(4), 232–242.

Oudshoorn, N., Rommes, E., & Stienstra, M. (2004). Configuring the user as everybody: Gender and design cultures in information and communication technologies. *Science, Technology & Human Values, 2* (1), 30–63.

Ozkal, K., Tekkaya, C., Cakiroglu, J., & Sungur, S. (in press). A conceptual model of relationships among constructivist learning environment perceptions, epistemological beliefs, and learning approaches. *Learning and Individual Differences.*

Papert, S. (1980). *Mindstorms: Children, computers, and powerful ideas.* Brighton, England: Harvester Press.

Papert, S. (1999). *Ghost in the machine: Seymour Papert on how computers fundamentally change the way kids learn.* Interview posted on ZineZone.com in 1999. Retrieved, March 1, 2008, from http://www.papert.org/articles/GhostsInTheMachine.html,

Papert, S. (2004). Keynote Address. *One to One Classroom Computing Conference.* May, Darling Harbour, Sydney, NSW.

Papert, S., & Cavillo, D. (2001). *The learning hub: Entry point into the twenty-first century learning.* A call for action at the local and global level. MIT Media Lab Future for Learning. Retrieved September 10, 2008, from http://www.papert.org

Parette, P., & McMahon, G. A. (2002). What should we expect of assistive technology? Being sensitive to family goals. *Teaching Exceptional Children, 35*(1), 56–61.

Parette, P., & Wojcik, B. W. (2004). Creating a technology toolkit for students with mental retardation: A systematic approach. *Journal of Special Education Technology, 19*(4), 23–31.

Paris, S., & Byrnes, J. (1992). The constructivist approach to self-regulation and learning in the classroom. *Journal of Experimental Child Psychology, 53*(1), 170–195.

Partners, H. (2005). *The emerging business of knowledge transfer: Creating value from intellectural products and services.* Report of a study commissioned by the Department of Education, Science and Training. Retrieved May 16, 2008, from http://www.dest.gov.au/NR/rdonlyres/75374026-6458-4D0E-9CF7-6D589645D093/4079/7253HERC05A1.pdf

Partners, H. (2006). *Changing paradigms: Rethinking innovation policies, practices and programs.* Report to the Business Council of Australia. Retrieved May 15, 2008 from http://www.howardpartners. com.au/publications/Changing_Paradigms_28_2_2006.pdf

Pascarella, P. (1997). Harnessing knowledge. *Management Review,* October, 37–40.

Perry, N., Hutchinson, L., & Thauberger, C. (2008). Talking about teaching self-regulated learning: Scaffolding student teachers' development and use of practices that promote self-regulated learning. *International Journal of Educational Research, 47,* 97–108.

Peterson-Karlan, G. R., & Parette, P. (2005). Millennial students with mild disabilities and emerging assistive technology trends. *Journal of Special Education Technology, 20*(4), 27–38.

Picot, A., & Wernick, C. (2007). The role of government in broadband access. *Telecommunications Policy, 31,* 660–674.

Pierson, M. & Howell, E. (2006). Pre-service teachers' perceptions of inclusion. *Academic Exchange, Fall,* 169–173.

Piscitello, L., & Rabbiosi, L. (2005). Reverse knowledge transfer: Organisational mechanisms and impact on the MNC performance. Preliminary evidence from the Italian case. *International Workshop on Innovation, Multinationals and Local Development,* September 30 – October 1, Catania, Italy. Retrieved June 1, 2008, from http://www.fscpo.unict.it/catania_workshop 2005/ Catania_workshop_2005/Piscitello%20&%20Rabbiosi%20(Abstract).pdf

Piscitello, L., & Rabbiosi, L. (2006). How does knowledge transfer from foreign subsidiaries affect parent companies' innovation capacity? *Proceedings of the DRUID Summer Conference on Knowledge, innovation and competitiveness: Dynamics of firms, networks, regions and institutions* (pp. 1–24). June 18 – 20, Copenhagen, Denmark. Retrieved June 1, 2008 from http://www. dime-eu.org/files/active/3/Rabbiosi.pdf

Polanyi, M. (1968). Logic and psychology. *American Psychologist, 23*(1), 27–43.

Prensky, M. (2001). *Digital game-based learning.* Columbus, OH: McGraw-Hill.

Prensky, M. (2005). In educational games, complexity matters. Mini-games are trivial – but complex games are not. *Educational Technology, 45*(4). Retrieved September 10, 2008, from http://www. marcprensky.com/writing/ Prensky_Complexity_matters.pdf.

Pridmore, T., Hilton, D., Green, J., Eastgate, R., & Cobb, S. (2004). *Mixed reality environments in stroke rehabilitation: Interfaces across the great real/virtual divide.* Paper presented at the 5th International Conference of Disability, Virtual Reality and Associated Technology, Oxford.

Prieger, J.E., & Hu, W.M. (2008). The broadband digital divide and the nexus of race, competition and quality. *Information Economics and Policy, 20,* 150–167.

Puckett, K. S. (2004). Project ACCESS: Field testing an assistive technology toolkit for students with mild disabilities. *Journal of Special Education Technology, 19*(2), 5–17.

Pumpa, M., Wyeld, T., & Adkins, B. (2006). Performing traditional knowledge using a game engine: Communicating and sharing Australian Aboriginal knowledge practices. *Proceedings of the 6th IEEE International Conference on Advanced Learning Technologies.* July 5–7, Kerkrade, The Netherlands

Queensland Government Office for Women. (2005). *Supporting women's participation in emerging industries: Science, engineering and technology.* Brisbane, QLD: Queensland Government. Retrieved March 12, 2007, from http://www.women.qld.gov.au/docs/Smart_Women_Smart_State/SET_concept_paper.pdf

Queensland Tertiary Admissions Centre (QTAC). (2006). *Overall positions.* Retrieved June 3, 2006, from http://www.qtac.edu.au/Year_10_and_11/ Overall_Positions.htm?activeMenuId=Information_For.Year_10_and_11&activeItemId=Information_For.Year_10_and_11.Overall_Positions_(OPs)

Rahim, A., & Golembiewski, R. T. (Eds.). (2005). *Current topics in management* (Vol. 10). Edison, NJ: Transition Publishers.

Randel, J., Morris, B., Wetzel, C., & Whitehill, B. (1992). The effectiveness of games for educational purposes: A review of recent research. *Simulation and Gaming, 23*(3), 261.

Raymond, E. (1999). *The Cathedral and the Bazaar.* O'Reilly, Sebastopol, CA.

Reading, C. (2007). *SiMERR National Website.* Retrieved March 1, 2008, from http://scs.une.edu.au/Web2/index.html

Reddick, A., Boucher, C. & Groseilliers, M. (2000). The dual digital divide: The information highway in Canada. Public Internet Advocacy Centre, Ottawa.

Richards, D., & Busch, D. (2005). Measuring, formalizing and modeling tacit knowledge. *Proceedings of International Conference on Artificial Intelligence and Application,* Innsbruck, Austria.

Riffel, L. A., Wehmeyer, M. L., Turnbull, A. P., Lattimore, J., Davies, D. K., Stock, S., et al. (2005). Promoting independent performance of transition-related tasks using a Palmtop PC-based self-directed visual and auditory prompting system. *Journal of Special Education Technology, 20*(2), 5–11.

Roberts, J. (2003). Trust and electronic knowledge transfer. *International Journal of Electronic Business, 1*(2), 168–186.

Robins, R., Dautenhahn, K., te Boekhorst, R., & Billard, A. (2004). Effects of repeated exposure to a humanoid robot on children with autism. In S. Keates, J. Clackson, P. Langdon & P. Robinson (Eds.), *Designing a more inclusive world,* London: Springer-Verlag, 225–236.

Roden, C. (1997). Young children's problem solving in design and technology: Towards a taxonomy of strategies. *The Journal of Design and Technology Education, 2* (1), 14–19.

Ronnback, S., Peikkari, J., Hyyppa, K., Berglund, T., & Koskinen, S. (2006). *A semi-autonomous wheelchair towards user-centred design.* Paper presented at the 10th International Conference on Computers Helping People with Special Needs, Linz, Austria.

Rose, D. (2000). Universal design for learning. *Journal of Special Education Technology, 15*(1), 67–70.

Rose, D. (2001). Universal design for learning. *Journal of Special Education Technology, 16*(4), 64–67.

Sachs, G. (2003). Dreaming with BRICS: The path to 2050. *Global economics paper no. 99.* Retrieved January 29, 2008 from http://www.gs.com/insight/research/reports/99.pdf

Samli, A. (Ed.). (1985). *Technology transfer: Geographic, economic, cultural and technical dimensions.* Westport, CT: Greenwood Press.

Scherer, M. J. (2005). *Living in the state of stuck: How assistive technology impacts the lives of people with disabilities* (4th ed.). Manchester, NH: Brookline Books.

Scherer, M. J., & Galvin, J. C. (1996). An outcomes perspective of quality pathways to the most appropriate technology. In J. C. Galvin & M. J. Scherer (Eds.), *Evaluating, selecting and using appropriate assistive technology,* Gaithersburg, MD: Aspen Publishers, 1–19.

Schwartz, A., Cui, X., Weber, D., & Moran, D. (2006). Brain-controlled interfaces: Movement restoration with neural prosthetics. *Neuron, 52,* 205–220.

Seale, J. (2006). Disability, technology and e-learning: Challenging conceptions. *Research in Learning Technology, 14*(1), 1–8.

Selamat, M. H., Abdullah, R., & Paul, C. J. (2006). Knowledge management framework in a technology support environment. *International Journal of Computer Science and Network Security, 6*(8A), 101–109.

Shannahan, K., Topping, K., & Bamford, J. (1994). Cross-school reciprocal peer tutoring of mathematics and makaton with children with severe learning difficulties. *British Journal of Learning Disabilities, 22*(3), 109–112.

Sheats, J. (2000). Information technology, sustainable development and developing nations. *Greener Management International, 32,* Winter, 33–41.

Shenk, D. (1997). *Data smog.* New York: Harper Edge.

Siach, T. (2004). *Governance and politics of China.* New York: Palgrave.

Singer, F. & Moscovici, H. (2008). Teaching and learning cycles in a constructivist approach to instruction. *Teaching and Teacher Education, 24,* 1613–1634.

Singh, N., Zhao, H., & Hu, X. (2005). Analyzing the cultural content of web sites: a cross-national comparison of China, India, Japan, and US. *International Marketing Review, 18,* 129.

Siperstein, G., Parker, R., Bardon, J., & Widaman, K. (2007). A national study of youth attitudes toward the inclusion of students with intellectual disabilities. *Exceptional Children, 73*(4), 435–445.

Skillen, M. (2008). Promoting thinking skills within the secondary classroom using digital media in IFIP Volume 281, *Learning to Live in a Knowledge Society.* Editors: E. Kendall & B. Samways, 126–199.

Smerdon, B., Cronen, S., Lanakar, L., Anderson, J., Iannotti, N., & Angeles, J. (2000). *Teachers tools for the twenty-first century.* Washington, DC: National Centre for Educational Statistics.

Smith, E. A. (2001). The role of tacit and explicit knowledge in the workplace. *Journal of Knowledge Management, 5*(4), 311–321.

Smith, P. J. (1999). So knowledge spillovers contribute to U.S. state output by growth? *Journal of Urban Economics, 45,* 331–353.

Smith, P. (2007). Have we made any progress? Including students with intellectual disabilities in regular education classrooms. *Journal of Intellectual and Developmental Disabilities, 45*(5), 297–309.

Smoot, S. (2004). An outcome measure for social goals of inclusion. *Rural Special Education Quarterly,* Summer, *23*(3), 15–22.

Smythe, P., Furner, S., & Mercinelli, M. (1995). Virtual reality technologies for people with special needs. In P. R. W. Roe (Ed.), *Telecommunications for all.* Brussels.

Spender, J. C. (1996). Making knowledge the basis of dynamic theory of the firm. *Strategic Management Journal, 17* (Winter), 45–62.

Spendlove, M. (2005). Knowledge transfer between SMEs & business schools: Opportunities and barriers. Paper presented at the 28th National Conference of the Institute for Small Business & Entrepreneurship, November. Retrieved June 1, 2008, from http://www.aston.ac.uk

Squire, K. (2005). *Changing the game: what happens when video games enter the classroom.* Accessed from: http://www.academiccolab.org/resources/documents/Changing%20The%20Game-final_2.pdf.

Squire, K., & Jenkins, H. (2003). Harnessing the power of games in education. *Insight, 3(5),* 5–33.

Stahl, S. (2003). Universal design for learning. *Journal of Special Education Technology, 18*(2), 65–67.

Stake, R. (1994). *The art of case study research.* Thousand Oaks, CA: Sage.

Standen, P., & Brown, D. (2005). *CyberPsychology & Behavior, 8* (3), 272–282.

Stenhoff, D., & Lignugaris/Kraft, B. (2007). A review of the effects of peer tutoring on students with mild disabilities in secondary settings. *Exceptional Children, 74*(1), 8–23.

Stenhoff, D., & Lignugaris-Kraft, B. (2007). A review of the effects of peer tutoring on students with mild disabilities in secondary settings. *Exceptional Children, 74* (1), 8–31.

Stoll, C. (1995). *Silicon snake oil: Second thoughts on the information highway.* New York: Anchor Books.

Straub, D., Loch, K., & Hill, C. (2001). Transfer of information technology to the Arab world: A test of cultural influence modeling. *Journal of Global Information Management, 9*(4), 6–28.

Strauss, A., & Corbin, J. (1998). *Basics of qualitative research* (2nd ed.). Thousand Oaks, CA: Sage.

Stringer, R., & Heath (2008). Academic self-perception and its relationship to academic performance. *Canadian Journal of Education, 31*(20). 327–345.

Sutton, R. (1991). Equity and computers in schools: A decade of research. *Review of Educational Research, 61*(4), 475-538.

Szczurek, M. (1982). *Meta-analysis of simulation games effectiveness for cognitive learning.* PhD dissertation, Indiana University.

Szecsi, T., & Giabo, D. (2007). International focus issue, *Childhood Education,* 338–341.

Tabb, L. (2008). A chicken in every pot; One laptop per child: The trouble with global campaign promises. *Journal of E-Learning, 5*(3).

Tassey, G. (2008). Globalisation of technology-based growth: The policy imperative. *Journal of Technology Transfer, 33,* 560–578.

Taub, E. (2008). New York Times. Retrieved June 1, 2008, from http://www.nytimes.com/2008/06/05/technology/personaltech/05basics.html?ref=personaltech

Technology Related Assistance for Individuals with Disabilities Act of 1988. (1988). Retrieved September 10, 2008, from www.section508.gov/docs/AT1998.html

Tesch, R. (1990). *Qualitative research: Analysis types and software tools.* New York: The Falmer Press.

Thorn, L. (2004). The Jhai Foundation. Retrieved June 4, 2006, from http://www.jhai.org

Timms, C., Courtney, L., & Anderson, N. (2006). Secondary girls' perceptions of advanced ICT subjects: Are they boring and irrelevant. *Australian Educational Computing, 21*(2), 3–8.

Timms, C., Lankshear, C., Anderson, N., & Courtney, L. (2008). Riding a hydra: Women ICT professionals' perceptions of working in the Australian ICT industry. *Information Technology and People,* 21(2), 155–177.

Tsang, E. W. (1995). The implementation of technology transfer in sino-foreign joint ventures. *International Journal of Technology Management, 10,* 757–766.

Turner, S.V., Bernt, P.W., & Pecora, N. (2002). *Why women choose information technology careers: Educational, social, and familial influences.* Paper presented at the Annual Meeting of the American Educational Research Association, New Orleans, Louisiana, April 2002. Retrieved January 9, 2006 from http://oak.cats.ohiou.edu/~turners/research/women.pdf

United Nations (UN). (2006). *List of member states.* Retrieved 8 June, 2008 from http://www.un.org/members/list.shtml

United States Congress (2004). The Individuals with Disabilities Education Improvement Act (IDEIA 2004). Washington D.C.: Author

Ure, J. Boucher, M., & Groseillers, M. (2000). *The digital divide: The information highway in Canada.* Public Interest Advocacy Centre, Ebo Research Associates.

Ure, J., Kuen, C. (1996). *Home computers and networking in Hong Kong.* Hong Kong University PC Survey.

U.S.–China Business Council. (2008). Foreign investment in China. Retrieved August 25, 2008, from http://www.uschina.org

Vandell, D., & Hembree, S. (1994). Peer social status and friendship: Independent contributors to children's social and academic adjustment. *Merrill-Palmer Quarterly, 49*(4), 461–477.

Vanderheiden, G. C. (1996). Computer access and use by people with disabilities. In J. C. Galvin & M. J. Scherer (Eds.), *Evaluating, selecting and using appropriate assistive technology.* Gaithersburg, MD: Aspen Publishers.

Van Dijk, J.A.G.M. (2006). Digital divide research, achievements and shortcomings, *Poetics, 34,* 221–235.

Van Dijk, J., & Hacker, K. (2003). The digital divide as a complex and dynamic phenomenon. *The Information Society,* 19, 315–326.

Van Eck, R. (2006). Digital game-based learning. *EDUCAUSE Review,* March–April, 17–30.

Van Sickle, R. (1986). A quantitative review of research on instructional simulation gaming: A twenty year perspective. *Theory and Research in Social Education, 14*(7), 245–264.

Vegso, J. (2005). Interest in CS as a major drops among incoming freshmen. *Computing Research News, 17*(3). Retrieved March 9, 2007 from http://www.cra.org/CRN/articles/may05/vegso.html

Velliste, M., Perel, S., Spalding, C., Whitford, S., & Schwartz, A. (2008). Cortical control of a prosthetic arm for self-feeding. *Nature*, May.

Ven, K. & Mannaert, H. (2008). Challenges and strategies in the use of open source software by independent vendors. *Information and Software Technology*, 50, 991–1002.

Victoria State Government. (2001). *Reality bytes: An in-depth analysis of attitudes about technology and career skills*. Victoria, Australia: Communications Division, Multimedia

von Hellens, L., & Nielsen, S. (2001). Australian women in IT. *Communications of the ACM, 44(7)*, 46–52.

von Hellens, L., Nielsen, S., & Trauth, E.M. (2001). *Breaking and entering the male domain. Women in the IT industry*. Proceedings of the 2001 ACM SIGCPR conference on computer personnel research. Special Interest Group on Computer Personnel Research Annual Conference. Retrieved January 9, 2006, from http://portal.acm.org/citation.cfm

von Krogh, G., & Roos, J. (1995). *Organizational epistemology*. London: Macmillan.

von Krogh, G., & Spaeth, S. (2007). The open source software phenomenon that promotes research. *Journal of Strategic Information Systems, 16*, 236–253.

Walters, K. (2006). EXI.T.Why women are shunning the technology industry. *Business Review Weekly*, July 27–August 2, 27–32.

Warren, M. (2007). The digital vicious cycle: Links between social disadvantage and digital exclusion in rural areas, *Telecommunications Policy, 31*, 374–388.

Watts, E. H., O'Brien, M., & Wojcik, B. W. (2004). Four models of assistive technology consideration: How do they compare to recommended educational assessment practices? *Journal of Special Education Technology, 19*(1), 43–56.

Wehmeyer, M. L. (1999). Assistive technology and students with mental retardation: Utilization and barriers. *Journal of Special Education Technology, 14*(1), 48–58.

Wehmeyer, M. L., Smith, S. J., Palmer, S. B., & Davies, D. K. (2004). Technology use by students with intellectual disabilities: An overview. *Journal of Special Education Technology, 19*(4), 7–21.

Wenglinski, H. (1998). *Does it compute? The relationship between educational technology and student achievement in mathematics*. Princeton, NJ: Policy Information Centre.

Wertheim, Margaret. (1999). *The pearly gates of cyberspace*. New York: W. W. Norton.

Wessels, R., Dijcks, B., Soede, M., Gelderblom, G. J., & De Witte, L. (2003). Non-use of provided assistive technology devices, a literature review. *Technology and Disability, 15*, 231–238.

West, M., & Ross, S. (2002). Retaining females in computer science: A new look at a persistent problem. *Journal of Computing in Small Colleges, 17*(5), 1–21.

Witte, S. (2007). First person "That's online writing not boring school writing": Writing with blogs and the talkback project. *Journal of Adolescent & Adult Literacy, 51*(2), 92–96.

Wong, D. (2008). Do contacts make a difference? The affects of mainstreaming on students' attitudes toward people with disabilities. *Research in Developmental Disabilities, 29*(1), January-February, 70-82.

Wood, L.E. (2008). Rural broadband: The provider matters. *Telecommunications Policy, 32,* 326–339.

Woodfield, R. (2002). Women and information systems development: Not just a pretty (inter)face? *Information Technology and People, 15*(2), 119–138.

Yan, A., & Gray, B. (1994). Bargaining power, management control, and performance in United States–China joint ventures: A comparative case study. *Academy of Management Journal, 37*(6), 1478–1517.

Yin, R. (2004). *Case Study Research: Design and Methods.* Sage Publications, London.

Yin, R. (2006). *Case study research: Design and methods* (3rd ed.). Thousand Oaks, CA: Sage Publications.

Yin, R. (2008). *Case study research: Design and methods* (4th ed.). Thousand Oaks, CA: Sage Publications.

Young, A. (1997). Higher-Order Learning and Thinking: What is it and How is it Taught? *Educational Technology,* July-August, 38-41

Young, J. (2003). *The extent to which information communication technology careers fulfill the career ideals of girls.* Proceedings of the 2003 Australian Women in IT Conference, Hobart, TAS, Australia.

Youngbae, K., Jeon, H., & Bae, S. (2008). Innovation patterns and policy implications of ADSL penetration in Korea: A case study. *Telecommunications Policy, 32,* 307–325.

Yu, X. Y. (1990). *International economic law.* Nanjing, China: University Press.

Zweben, S. (2005). 2003–2004 Taulbee Survey. Record Ph.D. production on the horizon; undergraduate enrolments continue in decline. *Computing Research News, 17*(3), Retrieved March 9, 2007, from http://www.cra.org/ CRN/articles/may05/taulbee.html

Zweben, S., & Aspray, W. (2004). 2001–2003 Taulbee Survey: Undergraduate enrollments drop: Department growth expectations moderate. *Computing Research News, 16*(3), 5–19.

Index

new
literacies
¶
AND DIGITAL EPISTEMOLOGIES

Colin Lankshear, Michele Knobel,
& Michael Peters
General Editors

New literacies and new knowledges are being invented "in the streets" as people from all walks of life wrestle with new technologies, shifting values, changing institutions, and new structures of personality and temperament emerging in a global informational age. These new literacies and ways of knowing remain absent from classrooms. Many education administrators, teachers, teacher educators, and academics seem largely unaware of them. Others actively oppose them. Yet, they increasingly shape the engagements and worlds of young people in societies like our own. The *New Literacies and Digital Epistemologies* series will explore this terrain with a view to informing educational theory and practice in constructively critical ways.

For further information about the series and submitting manuscripts, please contact:

Michele Knobel & Colin Lankshear
Montclair State University
Dept. of Education and Human Services
3173 University Hall
Montclair, NJ 07043
michele@coatepec.net

To order other books in this series, please contact our Customer Service Department at:

(800) 770-LANG (within the U.S.)
(212) 647-7706 (outside the U.S.)
(212) 647-7707 FAX

Or browse online by series at:

www.peterlang.com

www.ingramcontent.com/pod-product-compliance
Lightning Source LLC
Chambersburg PA
CBHW070943050326
40689CB00014B/3329